云计算工程师系列

云计算部署实战

主　编　肖　睿　吴振宇
副主编　杨智勇　胡文杰　赵红艳

中国水利水电出版社
www.waterpub.com.cn
·北京·

内 容 提 要

Openstack 目前获得了很多大公司的广泛支持，不仅能够搭建私有云，而且能够搭建公有云。本书针对具备 Linux 运维基础的人员，主要介绍了云计算平台 OpenStack、公有云 AWS、大数据 Hadoop 及 CDH 部署的相关知识与应用，以生动详细的理论讲解、企业级的实战项目案例，使读者能够掌握目前的热门技术。

本书通过通俗易懂的原理及深入浅出的案例，并配以完善的学习资源和支持服务，为读者带来全方位的学习体验，包括视频教程、案例素材下载、学习交流社区、讨论组等终身学习内容，更多技术支持请访问课工场 www.kgc.cn。

图书在版编目（CIP）数据

云计算部署实战 / 肖睿，吴振宇主编． -- 北京：
中国水利水电出版社，2017.5（2023.7 重印）
（云计算工程师系列）
ISBN 978-7-5170-5377-4

Ⅰ．①云… Ⅱ．①肖… ②吴… Ⅲ．①云计算 Ⅳ．
①TP393.027

中国版本图书馆CIP数据核字(2017)第099135号

策划编辑：祝智敏　　责任编辑：魏渊源　　加工编辑：高双春　　封面设计：梁 燕

书　名	云计算工程师系列 **云计算部署实战** YUNJISUAN BUSHU SHIZHAN
作　者	主　编　肖睿　吴振宇 副主编　杨智勇　胡文杰　赵红艳
出版发行	中国水利水电出版社 （北京市海淀区玉渊潭南路 1 号 D 座　100038） 网　址：www.waterpub.com.cn E-mail：mchannel@263.net（答疑） 　　　　sales@mwr.gov.cn 电　话：（010）68545888（营销中心）、82562819（组稿）
经　售	北京科水图书销售有限公司 电话：（010）68545874、63202643 全国各地新华书店和相关出版物销售网点
排　版	北京万水电子信息有限公司
印　刷	三河市德贤弘印务有限公司
规　格	184mm×260mm　16 开本　13.5 印张　292 千字
版　次	2017 年 5 月第 1 版　2023 年 7 月第 3 次印刷
印　数	6001—7000 册
定　价	39.00 元

凡购买我社图书，如有缺页、倒页、脱页的，本社营销中心负责调换

版权所有·侵权必究

丛书编委会

主　任：肖　睿

副主任：刁景涛

委　员：杨　欢　　潘贞玉　　张德平　　相洪波　　谢伟民
　　　　庞国广　　张惠军　　段永华　　李　娜　　孙　苹
　　　　董泰森　　曾谆谆　　王俊鑫　　俞　俊

课工场：李超阳　　祁春鹏　　祁　龙　　滕传雨　　尚永祯
　　　　张雪妮　　吴宇迪　　曹紫涵　　吉志星　　胡杨柳依
　　　　李晓川　　黄　斌　　宗　娜　　陈　璇　　王博君
　　　　刁志星　　孙　敏　　张　智　　董文治　　霍荣慧
　　　　刘景元　　袁娇娇　　李　红　　孙正哲　　史爱鑫
　　　　周士昆　　傅　峥　　于学杰　　何娅玲　　王宗娟

前　　言

"互联网 + 人工智能"时代，新技术的发展可谓是一日千里，云计算、大数据、物联网、区块链、虚拟现实、机器学习、深度学习等等，已经形成一波新的科技浪潮。以云计算为例，国内云计算市场的蛋糕正变得越来越诱人，以下列举了 2016 年以来发生的部分大事。

1. 中国联通发布云计算策略，并同步发起成立"中国联通沃云 + 云生态联盟"，全面开启云服务新时代。

2. 内蒙古斥资 500 亿元欲打造亚洲最大云计算数据中心。

3. 腾讯云升级为平台级战略，旨在探索云上生态，实现全面开放，构建可信赖的云生态体系。

4. 百度正式发布"云计算 + 大数据 + 人工智能"三位一体的云战略。

5. 亚马逊 AWS 和北京光环新网科技股份有限公司联合宣布：由光环新网负责运营的 AWS 中国（北京）区域在中国正式商用。

6. 来自 Forrester 的报告认为，AWS 和 OpenStack 是公有云和私有云事实上的标准。

7. 网易正式推出"网易云"。网易将先行投入数十亿人民币，发力云计算领域。

8. 金山云重磅发布"大米"云主机，这是一款专为创业者而生的性能王云主机，采用自建 11 线 BGP 全覆盖以及 VPC 私有网络，全方位保障数据安全。

DT 时代，企业对传统 IT 架构的需求减弱，不少传统 IT 企业的技术人员，面临失业风险。全球最知名的职业社交平台 LinkedIn 发布报告，最受雇主青睐的十大职业技能中"云计算"名列前茅。2016 年，中国企业云服务整体市场规模超 500 亿元，预计未来几年仍将保持约 30% 的年复合增长率。未来 5 年，整个社会对云计算人才的需求缺口将高达 130 万。从传统的 IT 工程师转型为云计算与大数据专家，已经成为一种趋势。

基于云计算这样的大环境，课工场（kgc.cn）的教研团队几年前开始策划的"云计算工程师系列"教材应运而生，它旨在帮助读者朋友快速成长为符合企业需求的、优秀的云计算工程师。这套教材是目前业界最全面、专业的云计算课程体系，能够满足企业对高级复合型人才的要求。参与本书编写的院校老师还有吴振宇、杨智勇、胡文杰赵红艳等。

课工场是北京大学下属企业北京课工场教育科技有限公司推出的互联网教育平台，专注于互联网企业各岗位人才的培养。平台汇聚了数百位来自知名培训机构、高校的顶级名师和互联网企业的行业专家，面向大学生以及需要"充电"的在职人员，针对与互联网相关的产品设计、开发、运维、推广和运营等岗位，提供在线的直播和录播课程，并通过遍及全国的几十家线下服务中心提供现场面授以及多种形式的教学服务，并同步研发出版最新的课程教材。

除了教材之外，课工场还提供各种学习资源和支持，包括：
- 现场面授课程
- 在线直播课程
- 录播视频课程
- 授课 PPT 课件
- 案例素材下载
- 扩展资料提供
- 学习交流社区
- QQ 讨论组（技术，就业，生活）

以上资源请访问课工场网站 www.kgc.cn。

本套教材特点

（1）科学的训练模式
- 科学的课程体系。
- 创新的教学模式。
- 技能人脉，实现多方位就业。
- 随需而变，支持终身学习。

（2）企业实战项目驱动
- 覆盖企业各项业务所需的 IT 技能。
- 几十个实训项目，快速积累一线实践经验。

（3）便捷的学习体验
- 提供二维码扫描，可以观看相关视频讲解和扩展资料等知识服务。
- 课工场开辟教材配套版块，提供素材下载、学习社区等丰富的在线学习资源。

读者对象

（1）初学者：本套教材将帮助你快速进入云计算及运维开发行业，从零开始逐步成长为专业的云计算及运维开发工程师。

（2）初中级运维及运维开发者：本套教材将带你进行全面、系统的云计算及运维开发学习，逐步成长为高级云计算及运维开发工程师。

课程设计说明

课程目标

读者学完本书后,能够掌握云计算、大数据的相关原理,了解云计算数据中心,能够完成 OpenStack 体系架构、大数据体系架构的设计与部署。

训练技能

- 理解云计算相关概念及 OpenStack 各个组件功能。
- 掌握 OpenStack 云平台的部署。
- 理解大数据 Hadoop 原理并掌握其应用,能够使用 CDH 部署。
- 掌握公有云 AWS 的服务架构及应用。

设计思路

本书采用了教材 + 扩展知识的设计思路,扩展知识提供二维码扫描,形式可以是文档、视频等,内容可以随时更新,能够更好地服务读者。

教材分为 8 个章节、3 个阶段来设计学习,即云计算、大数据、云计算数据中心,具体安排如下:

- 第 1 章~第 4 章介绍云计算相关基础知识,理解 OpenStack 系统架构与各个组件功能,掌握 OpenStack 云平台的部署过程。
- 第 5 章~第 7 章介绍大数据相关基础知识,理解 Hadoop 架构组成、HDFS、MapReduce 架构、HBase 相关概念,掌握 CDH 部署过程。
- 第 8 章介绍云计算数据中心与公有云 AWS,理解云计算数据中心相关核心概念与云计算数据中心体系架构,掌握公有云 AWS 的服务架构及应用。

章节导读

- 技能目标:学习本章所要达到的技能,可以作为检验学习效果的标准。
- 本章导读:对本章涉及的技能内容进行分析并展开讲解。
- 操作案例:对所学内容的实操训练。
- 本章总结:针对本章内容的概括和总结。
- 本章作业:针对本章内容的补充练习,用于加强对技能的理解和运用。
- 扩展知识:针对本章内容的扩展、补充,对于新知识随时可以更新。

学习资源

- 学习交流社区(课工场)
- 案例素材下载
- 相关视频教程

更多内容详见课工场 www.kgc.cn。

目　录

前言
课程设计说明

第 1 章　OpenStack 体验 1
1.1　云计算介绍 2
1.1.1　红帽 OpenStack 产品 2
1.1.2　安装红帽 OpenStack 环境 3
1.2　创建实例案例 7
1.2.1　准备工作 8
1.2.2　在 Horizon 中启动实例 10
1.2.3　扩展应用 23
本章总结 26
本章作业 26

第 2 章　OpenStack 搭建企业私有云 27
2.1　环境准备 28
2.2　基础配置 32
2.2.1　安装配置 NTP 服务 32
2.2.2　配置 OpenStack yum 库 33
2.2.3　MySQL 数据库 34
2.2.4　NoSQL 数据库 37
2.2.5　安装配置 Messaging server-RabbitMQ 37
2.2.6　Memcached 39
2.3　认证服务 39
2.3.1　认证服务概览 39
2.3.2　安装配置 40
2.3.3　创建服务实体和 API 端点 42
2.3.4　创建域、项目、用户和角色 44
2.3.5　验证操作 47
2.3.6　创建 OpenStack 客户端环境脚本 ... 48
2.4　镜像服务 49
2.4.1　镜像服务概览 49
2.4.2　安装和配置 50
2.4.3　创建镜像服务的 API 端点 51
2.4.4　安装软件包 52
2.4.5　验证操作 53
2.5　计算服务 55
2.5.1　计算服务概览 55
2.5.2　安装并配置控制节点 56
2.5.3　安装和配置计算节点 60
2.5.4　验证操作 62
2.6　Networking 服务 63
2.6.1　网络服务概览 63
2.6.2　安装并配置控制节点 64
2.6.3　安装和配置计算节点 70
2.6.4　验证操作 72
2.7　Dashboard 74
2.8　启动一个实例 76
本章总结 83

第 3 章　云存储 85
3.1　块存储与文件存储 86
3.2　对象存储 87
3.3　对象存储 Swift 88
3.3.1　Swift 数据模型 88
3.3.2　Swift 组件 89
3.3.3　Swift 的数据一致性 91
3.3.4　Swift 存储策略 91
3.3.5　对象存储 Swift 补充 92
3.4　块存储服务 Cinder 94
3.4.1　块存储服务概览 94
3.4.2　块存储服务组件 94
3.4.3　Cinder 架构解释 95
3.4.4　Cinder 支持存储类型 96
本章总结 96

第 4 章　Fuel 安装 OpenStack 97
　4.1　Fuel 概述 98
　4.2　虚拟环境设置 100
　4.3　部署 OpenStack 环境 105
　本章总结 116

第 5 章　大数据 Hadoop 117
　5.1　什么是大数据 118
　5.2　Hadoop 体系结构 119
　5.3　安装 Hadoop 运行环境 121
　　5.3.1　在 Linux 中配置 Hadoop 运行环境 . 121
　　5.3.2　Hadoop 完全分布式安装 124
　　5.3.3　运行 Hadoop 的 WordCount 程序 .. 130
　5.4　HDFS 体系结构 132
　　5.4.1　基本概念 133
　　5.4.2　Master/Slave 架构 136
　　5.4.3　HDFS 的 Web 界面 136
　　5.4.4　HDFS 的命令行操作 138
　5.5　MapReduce 基础 141
　　5.5.1　MapReduce 概述 142
　　5.5.2　MapReduce 架构设计 142
　　5.5.3　MapReduce 编程模型 143
　5.6　下一代 MapReduce 框架 YARN 145
　　5.6.1　YARN 架构 145
　　5.6.2　YARN 配置文件 146
　　5.6.3　YARN 作业执行流程 147
　　5.6.4　YARN 优势 148
　本章总结 148
　本章作业 149

第 6 章　HBase 数据库 151
　6.1　HBase 基础 152
　　6.1.1　HBase 简介 152

　　6.1.2　HBase 体系结构 153
　　6.1.3　HBase 数据模型 155
　　6.1.4　HBase 的安装 158
　6.2　HBase Shell 操作 165
　6.3　MapReduce 与 HBase 172
　6.4　Hive 和 Spark 174
　　6.4.1　Hive 174
　　6.4.2　Spark 175
　本章总结 176
　本章作业 176

第 7 章　部署 CDH 环境 177
　7.1　CDH 概述 178
　7.2　案例环境 178
　　7.2.1　准备工作 179
　　7.2.2　安装数据库 182
　　7.2.3　安装 CDH 183
　　7.2.4　安装配置 CDH 集群 185
　　7.2.5　配置 Kafka 190
　本章总结 192

第 8 章　云计算数据中心与亚马逊 AWS 193
　8.1　体系结构简介 194
　8.2　云计算数据中心特点 195
　8.3　网络应用架构 196
　8.4　能源利用 199
　8.5　自动化管理 201
　8.6　容灾系统 201
　　8.6.1　容灾概述 202
　　8.6.2　容灾技术 202
　8.7　亚马逊 AWS 203
　本章总结 205

第 1 章

OpenStack 体验

技能目标

- 理解云计算核心概念
- 理解 OpenStack 各组件作用
- 会使用 Packstack 部署 OpenStack 环境
- 会使用 Dashboard 启动 OpenStack 实例

本章导读

云计算是一种模型，能够提供无论在何时何地都可以便捷获取所需资源的模型，这些资源可以是网络资源、存储资源、服务器资源、甚至是服务或者应用软件资源等，这些资源能够让用户根据需要快速创建应用，并且在不需要时进行资源释放。

红帽 Openstack 产品 RHEL OSP 一方面负责与运行在物理节点上的 Hypervisor 进行交互，实现对各种硬件资源的管理与控制，另一方面为用户提供一个满足要求的虚拟实例。

知识服务

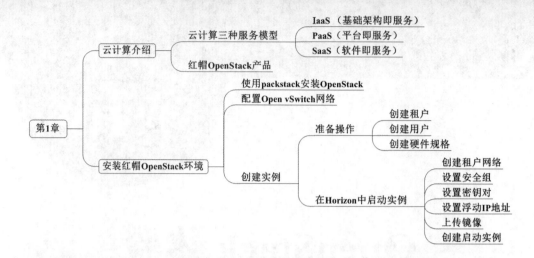

1.1 云计算介绍

云计算是一种模型,能够提供无论在何时何地都可以便捷获取所需资源的模型,这些资源可以是网络资源、存储资源、服务器资源,甚至是服务或者应用软件资源等,这些资源能够让用户根据需要快速创建应用,并且在不需要时进行资源释放。

云计算模型中有三种基本服务模型可用:

1. IaaS

IaaS(基础架构即服务):提供最底层的IT基础设施服务,包括处理能力、存储空间、网络资源等,用户可以从中获取硬件或者虚拟硬件资源(包括裸机或者虚拟机),然后可以给申请到的资源安装操作系统和其他应用程序。一般面向对象是IT管理人员。

2. PaaS

PaaS(平台即服务):把已经安装好开发环境的系统平台作为一种服务通过互联网提供给用户。用户可以在上面安装其他应用程序,但不能修改已经安装好的操作系统和运行环境。一般面向对象是开发人员,需要了解平台提供环境下的应用开发和部署。

3. SaaS

SaaS(软件即服务):可直接通过互联网为用户提供软件和应用程序的服务。用户可以通过租赁的方式获取安装在厂商或者服务供应商那里的软件。一般面向的对象是普通用户,最常见模式就是提供给用户一组账号和密码。

1.1.1 红帽 OpenStack 产品

红帽提供了很多和整合云计算技术相关的产品,其中 RHEL OSP(RedHat Enterprise Linux OpenStack Platform)是红帽一款扩展性极强的 IaaS 产品。红帽的这款 OpenStack 产品一方面负责与运行在物理节点上的 Hypervisor 进行交互,实现对各种硬件资源的管理与控制，另一方面为用户提供一个满足要求的虚拟实例。

OpenStack 项目主要提供：计算服务、存储服务、镜像服务、网络服务，均依赖于身份认证 keystone 的支撑。其中的每个项目可以拆开部署，同一项目也可以部署在多台物理机上，并且每个服务都提供了应用接口程序（API），方便与第三方集成调用资源。OpenStack 服务如表 1-1 所示。

表 1-1 OpenStack 服务

服务	项目名称	描述
Compute（计算服务）	Nova	负责实例生命周期的管理，计算资源的单位。对 Hypervisor 进行屏蔽，支持多种虚拟化技术（红帽默认为 KVM），支持横向扩展
Network（网络服务）	Neutron	负责虚拟网络的管理，为实例创建网络的拓扑结构。是面向租户的网络管理，可以自己定义自己的网络，各个租户之间互不影响
Identity（身份认证服务）	Keystone	类似于 LDAP 服务，对用户、租户和角色、服务进行认证与授权，并且支持多认证机制
Dashboard（控制面板服务）	Horizon	提供一个 Web 管理界面，与 OpenStack 底层服务进行交互
Image Service（镜像服务）	Glance	提供虚拟机镜像模板的注册与管理，将做好的操作系统复制为镜像模板，在创建虚拟机时直接使用。可支持多格式的镜像
Block Storage（块存储服务）	Cinder	负责为运行实例提供持久的块存储设备，可进行方便的扩展，按需付费，支持多种后端存储
Object Storage（对象存储服务）	Swift	为 OpenStack 提供基于云的弹性存储，支持集群无单点故障
Telemetry（计量服务）	Ceilometer	用于度量、监控和控制数据资源的集中来源，为 OpenStack 用户提供记账途径

1.1.2 安装红帽 OpenStack 环境

安装红帽 OpenStack 环境的硬件需求如表 1-2 所示。

表 1-2 硬件要求

硬件 \ 需求	云控制节点	计算节点
CPU	支持 Intel 64 或 AMD64 CPU 扩展，并启用了 AMD-V 或 Intel VT 硬件虚拟化支持的 64 位 x86 处理器	
内存	2GB	至少 2GB
磁盘空间	50GB	最低 50GB
网络	2 个 1Gbps 网卡	2 个 1Gbps 网卡

本案例使用三台主机，三台主机的系统均采用 RHEL7.1 操作系统，其中 servera、serverb 主机采用最小化安装操作系统，workstation 主机安装桌面环境，主体操作将在 workstation 主机上完成。在 servera、serverb 主机的 hosts 文件中添加 IP 地址解析记录。网络设计为：公有网段采用 172.25.18.0/24（保证此网段可以访问 Internet），私有网段采用 172.24.18.0/24。

具体案例环境如表 1-3 所示。

表 1-3　案例环境

计算机名	IP 地址	角色
workstation.example.com	eth0: 172.25.18.9 eth1: 172.24.18.9	不安装 OpenStack 组件，作为配置 OpenStack 的客户端
servera.example.com	eth0: 172.25.18.10 eth1:172.24.18.10	第一台服务器，安装回答文件中指定的组件（默认全部）。作为专用云控制节点，网络节点，存储节点，计算节点
serverb.example.com	eth0: 172.25.18.11 eth1:172.24.18.11	第二台服务器，扩展计算节点（还可以再安装多台计算节点）

红帽 OpenStack 是使用运维自动化工具 Puppet 模块来便利部署和管理的，需要安装 openstack-packstack 软件包，使用 packstack 生成回答文件，根据管理员需求来配置修改回答文件，之后将回答文件传递到安装程序来完成 OpenStack 的自动安装部署。

1. 使用 packstack 安装 OpenStack

```
[root@servera ~]#yum install -y openstack-packstack
```

使用 packstack 结合 --gen-answer-file 选项来生成默认的回答文件。

```
[root@servera ~]# packstack --gen-answer-file /root/answer.txt
Packstack changed given value  to required value /root/.ssh/id_rsa.pub
```

通过修改生成的回答文件，来定义自己的 OpenStack 环境。

```
[root@servera ~]#vim /root/answer.txt
  CONFIG_DEFAULT_PASSWORD=redhat
  CONFIG_NTP_SERVERS=pool.ntp.org        //NTP 服务器同步时间（这里是官方的 NTP
    服务器，如果是不能上网的实验环境需要自行安装 NTP 服务器）
  CONFIG_KEYSTONE_ADMIN_PW=redhat        //keystone admin 初始密码
  CONFIG_NEUTRON_OVS_TUNNEL_IF=eth1      // 这里使用的是默认的 VXLAN 网络，eth1 是
    默认的内网网卡，之后需要我们配置 br-ex 作为外部路由的默认桥
  CONFIG_HORIZON_SSL=y                   // 为 Horizon 开启 SSL
  CONFIG_PROVISION_DEMO=n                // 不使用演示的用法，不配置 demo 用户
  CONFIG_COMPUTE_HOSTS=172.25.18.10,172.25.18.11   // 计算节点主机
```

如果将不同的主机安装不同的角色，只需要分别在 CONFIG_COMPUTE_HOSTS 后填写控制节点 IP 地址，CONFIG_COMPUTE_HOSTS 后填写计算节点 IP 地址，

CONFIG_NETWORK_HOSTS 后添加网络节点 IP 地址，一般控制节点设置为一台主机，计算节点和网络节点设置为使用逗号分隔的多台主机。

使用 packstack 结合 --answer-file 选项来使用 Puppet 按照回答文件中定义的内容来自动部署 OpenStack 环境。

```
[root@servera ~]#packstack --answer-file /root/answer.txt
Welcome to the Packstack setup utility
The installation log file is available at: /var/tmp/packstack/20160218-103723-ZPbrHL/openstack-setup.log

Installing:
Clean Up                                     [ DONE ]
root@172.25.18.11's password:
root@172.25.18.10's password:
Setting up ssh keys                          [ DONE ]
Discovering hosts' details                   [ DONE ]
Adding pre install manifest entries          [ DONE ]
Installing time synchronization via NTP      [ DONE ]
……
……
Finalizing                                   [ DONE ]
 **** Installation completed successfully ******
Additional information:
 * File /root/keystonerc_admin has been created on OpenStack client host 172.25.0.10. To use the
   command line tools you need to source the file.
 * NOTE : A certificate was generated to be used for ssl, You should change the ssl certificate
   configured in /etc/httpd/conf.d/ssl.conf on 172.25.0.10 to use a CA signed cert.
 * To access the OpenStack Dashboard browse to https://172.25.18.10/dashboard.
Please, find your login credentials stored in the keystonerc_admin in your home directory.
 * To use Nagios, browse to http://172.25.18.10/nagios username: nagiosadmin, password:
   13bf55c7bdd7442c
 * The installation log file is available at: /var/tmp/packstack/20160218-110841-Te7Vjd/openstack-
   setup.log
 * The generated manifests are available at: /var/tmp/packstack/20160218-110841-Te7Vjd/manifests
```

OpenStack 环境部署完成后，使用 "https:// 控制节点 IP 地址 /dashboard" 的方式登录 OpenStack 的 Horizon Web 界面，登录页面如图 1.1 所示。

```
[root@workstation ~]# firefox https://172.25.18.10/dashboard
```

2. 配置 Open vSwitch 网络

为了使 OpenStack 的虚拟网络与真实网络进行连通，需要配置 Open vSwitch。Open vSwitch 是开放虚拟交换标准，是虚拟平台上通过软件方式形成的交换机部件，即虚拟交换机，支持标准的接口和协议，其数量和端口数目都可以灵活配置。

类似于桥接，Open vSwitch 定义两个接口，br-int 用于连接内部网络，br-ex 用于连接外部网络，eth0 作为 br-ex 上的一个接口。通过网卡配置文件配置 Open vSwitch。

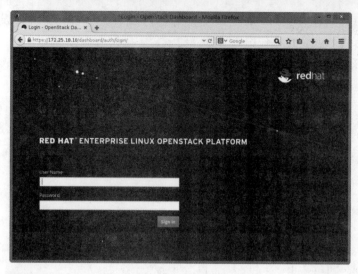

图 1.1 Horizon Web 登录界面

```
[root@servera ~]# cd /etc/sysconfig/network-scripts/
[root@servera network-scripts]# cp ifcfg-eth0 ifcfg-br-ex
[root@servera network-scripts]# vim ifcfg-br-ex
DEVICE=br-ex
BOOTPROTO=static
ONBOOT=yes
TYPE=OVSBridge
DEVICETYPE=ovs
IPADDR=172.25.18.10
NETMASK=255.255.255.0
GATEWAY=172.25.18.254
DNS1=8.8.8.8

[root@servera network-scripts]#vim ifcfg-eth0
DEVICE=eth0
ONBOOT=yes
TYPE=OVSPort
OVS_BRIDGE=br-ex
DEVICETYPE=ovs
[root@servera network-scripts]# systemctl restart network
```

查看 br-ex 详细信息，IP 地址已经配置完毕。

```
[root@servera network-scripts]# ifconfig br-ex
br-ex: flags=4163<UP,BROADCAST,RUNNING,MULTICAST>mtu 1500
   inet 172.25.18.10  netmask 255.255.255.0  broadcast 172.25.18.255
      inet6 fe80::496:b8ff:fe72:8649prefixlen 64  scopeid 0x20<link>
ether 06:96:b8:72:86:49  txqueuelen 0  (Ethernet)
      RX packets 1252  bytes 124882 (121.9 KiB)
      RX errors 0  dropped 0  overruns 0  frame 0
```

TX packets 1116 bytes 123163 (120.2 KiB)
TX errors 0 dropped 0 overruns 0 carrier 0 collisions 0

3. 登录 OpenStack Web 管理平台

Horizon Web 界面可供操作员和管理员管理其 OpenStack 平台。Horizon 与每个 OpenStack 项目 API 进行通信，可以使用此 Web 界面执行大部分任务。登录使用的用户和密码可以查看安装 OpenStack 过程中生成的 keystone_admin 文件。

```
[root@servera ~]# cat /root/keystonerc_admin
export OS_USERNAME=admin
export OS_TENANT_NAME=admin
export OS_PASSWORD=redhat
export OS_AUTH_URL=http://172.25.18.10:5000/v2.0/
export OS_REGION_NAME=RegionOne
export PS1='[\u@\h \W(keystone_admin)]\$ '
```

使用 Admin 用户登录 Horizon Web 界面如图 1.2 所示。

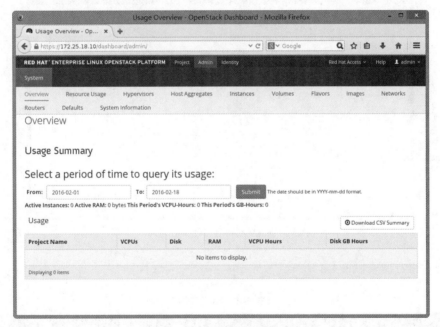

图 1.2 Admin 登录页面

通过修改 Horizon Web 的配置文件 /etc/openstack-dashboard/local_settings 定义可以登录到 Horizon Web 的主机，默认为 ALLOWED_HOSTS = ['*',] 表示所有主机均可登录。

1.2 创建实例案例

案例需求：通过 Horizon Web 前端创建符合要求的实例，并添加相应规则允许用户可以使用 SSH 协议登录实例，以及可以访问实例的 80 端口（安全组内访问）及 443 端口（外网访问）。

1.2.1 准备工作

1. 创建租户

在 OpenStack 中租户是对一个项目的描述，其中包含已分配数量的 OpenStack 用户和资源。创建租户时就先配置了一组资源配额，该配额包含该项目总共可以使用的 VCPU、实例、RAM 以及分配给实例的浮动 IP 等资源的数量。租户信息在 Identity 选项卡下的 Projects 子选项卡中查看，如图 1.3 所示。

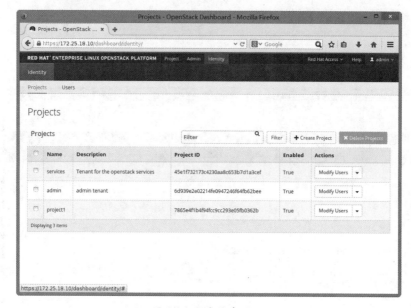

图 1.3　租户信息页面

创建租户需求如表 1-4 所示。

表 1-4　租户参数

参数	值
租户名称	Project1
配额	4 个 VCPU、2 个实例、4096RAM、2 个浮动 IP

在图 1.3 所示的租户信息页面中，单击 Create Project 按钮按照表 1-4 内容来创建租户，其中创建的配额信息如图 1.4 所示。

2. 创建用户

创建完租户后，需要创建一个用户来管理租户，用户信息如表 1-5 所示。

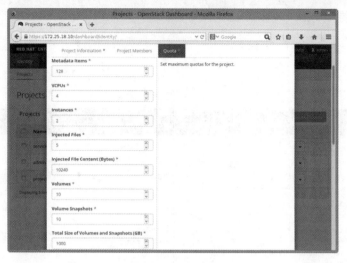

图 1.4　创建租户

表 1-5　用户参数

参数	值
用户名	user1
电子邮件	user@123.com
密码	redhat
主项目	project1
角色	_member_

在 Identity 选项卡下的 Users 子选项卡中，单击 Create User 按钮来创建用户，如图 1.5 所示。

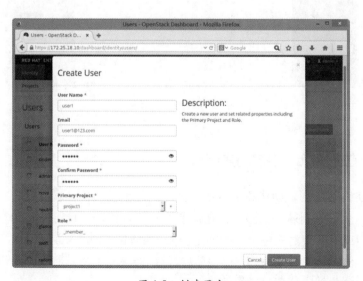

图 1.5　创建用户

3. 创建硬件规格

在创建实例之前，需要创建一个硬件规格供之后创建实例时使用，硬件规格参数如表 1-6 所示。

表 1-6　硬件规格参数

参数	值
名称	M1.small
ID	auto
VCPU	1
RAM MB	2048
Root Disk GB	20
Ephemeral Disk GB	0
Swap Disk MB	512

在 Admin 选项卡下的 System 子选项卡中选择 Flavors，单击 Create Flavor 按钮来创建硬件规格，如图 1.6 所示。

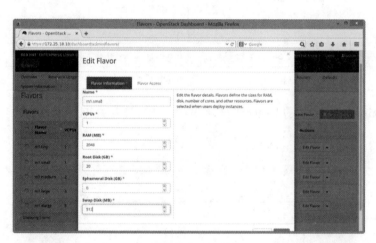

图 1.6　创建硬件规格

1.2.2　在 Horizon 中启动实例

在启动实例前，需要创建和上传镜像文件，此镜像文件包含可以使用的操作系统甚至可以是安装好常用软件的系统；还需要配置安全组，以便打开实例与防火墙之间的某些端口，或者实现对实例上运行特定服务的访问；分配浮动 IP 地址，以便外网用户可以访问实例；创建 SSH 密钥对，可以方便地登录实例进行管理与使用。这些操作可以直接使用之前创建管理租户的用户来完成。

使用 user1 用户登录 dashboard 管理租户 project1，如图 1.7 所示。

图 1.7　使用 user1 登录配置

1. 创建网络

OpenStack 网络是由 Neutron 项目负责，利用灵活的物理网络划分机制，在多租户的环境下提供给每个租户独立的网络环境。这个"网络"是一个可以被用户创建的对象，相当于物理环境下的交换机，有多个可以被动态创建和销毁的虚拟端口。Neutron 中的路由器也是一个路由选择和准发的虚拟设备，可以被创建和销毁的部件。OpenStack 中网络配置相关参数如表 1-7 所示。

表 1-7　网络配置参数

参数	值
专用网络名称	Private
专用子网信息	子网名称：subnet1 网络地址：192.168.0.0/24（可以自定义，也可以用 172.24.18.0/24）
子网详细信息	启用 DHCP，给虚拟实例自动分配私有地址使用，可定义地址池分配范围
公共网络名称	Public
公共子网信息	子网名称：subnet2 网络地址：172.25.18.0/24 网关：172.25.18.254
公共子网详细信息	不启动 DHCP，可定义浮动 IP 地址范围（如 172.25.18.25 ～ 172.25.18.99）
路由器名称	Router1
外部网络	Public
路由器网关	Public

在 Project 选项卡的 Network 子选项卡中选择 Networks，单击 Create Network 按钮来创建虚拟网络，包含私网和公网两个子网。

根据表 1-7 填写私有网络信息，创建私有网络，如图 1.8 所示。

图 1.8　私有网络设置

根据公有网络信息创建公有网络，如图 1.9 所示。

图 1.9　公有网络设置

其中公有网络还需要使用管理员账号 Admin 定义为外部网络，在 Admin 选项卡下的 System 子选项卡中选择 Networks，选择定义的公有网络单击 Edit Network 按钮进行编辑，勾选 External Network，定义为外网，如图 1.10 所示。

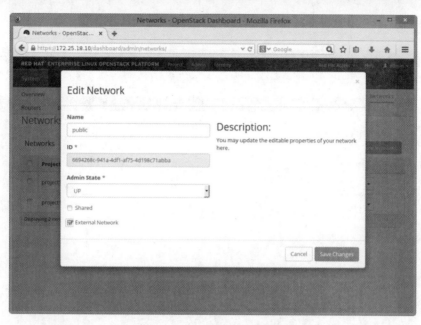

图 1.10　管理员定义外部网络

接下来继续使用 user1 账户，在 Project 选项卡下的 Network 子选项卡中选择 Routers，单击 Create Router 按钮创建虚拟路由器，如图 1.11 所示。

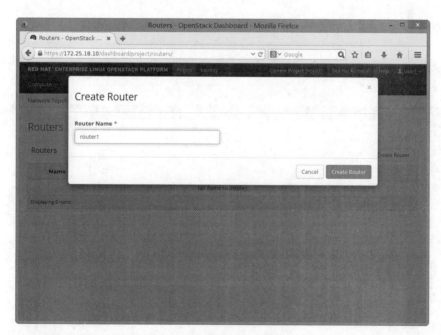

图 1.11　创建虚拟路由器

Neutron 中的虚拟路由器一般有一个网关（Gateway）用于连接外网，可以有多个接口（Interface）用于连接租户网络的子网。

设置路由器的网关为公网 public 网段的 IP 地址，如图 1.12 所示。

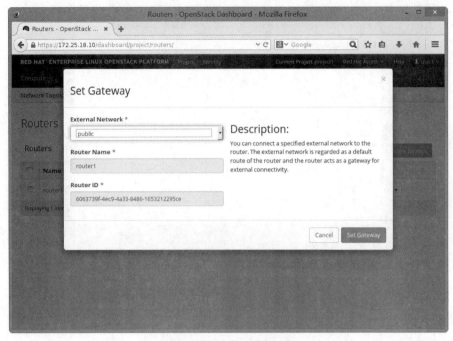

图 1.12　创建路由器网关

添加路由器接口为 private 的网段 IP 地址，如图 1.13 所示。

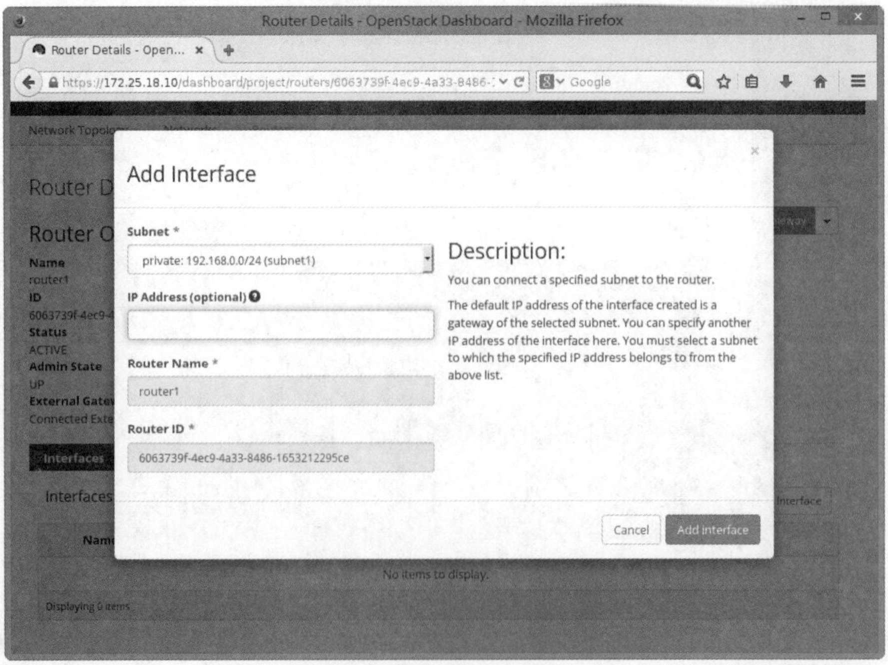

图 1.13　创建路由器接口

没有指定接口的具体 IP 地址，会获得第一个 private 设定的 IP 地址，如图 1.14 所示。

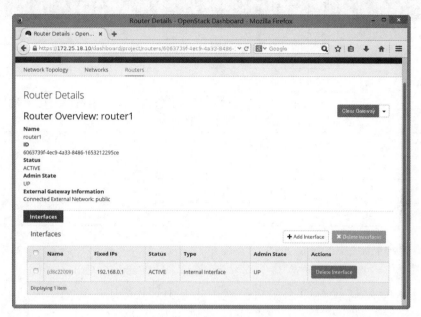

图 1.14　路由器接口信息

OpenStack 网络环境创建完成后，可以在 Project 选项卡下的 Network 子选项卡中的 Network Topology 中查看网络拓扑，如图 1.15 所示。

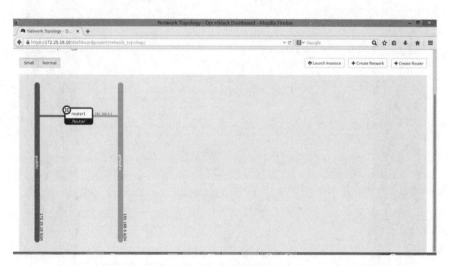

图 1.15　网络拓扑

2．访问和安全

在 Project 选项卡下的 Compute 子选项卡中的 Access&Security 选项中可以分别定义 Security Groups（安全组）、Key Pairs（密钥对）以及 Floating IPs（浮动 IP 地址），相关参数如表 1-8 所示。

表 1-8 访问和安全参数

参数	值
为实例分配的 IP	172.25.18.26
安全组名称	Sec1
安全组描述	Web 和 SSH
安全组权限	允许来自 0.0.0.0/0 的 SSH（TCP/22）和 HTTPS（TCP/443）以及来自 Sec1 组的 HTTP（TCP/80）
SSH 密钥对名称	Key1
映像名称	Image1
映像格式	QCOW2——QEMU 模拟器
映像设置	磁盘无最低限制，内存最低为 1024MB RAM

（1）创建安全组

设置安全组，以便开放某些防火墙端口。在 Security Groups 中，单击 Create Security Group 来创建新的安全组，分别添加 HTTP 规则，如图 1.16 所示；添加 HTTPS 规则，如图 1.17 所示；添加 SSH 规则，如图 1.18 所示。即分别开放防火墙的 80、443、22 端口，供不同用户访问连接。

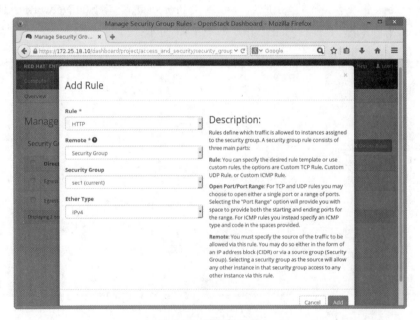

图 1.16　HTTP 规则

（2）创建密钥对

在 Key Pairs 选项卡中，单击 Create Key Pair 来创建密钥对，并将密钥保存到 workstation 主机本地，方便用户可以不需要输入密码直接登录实例。创建密钥如图 1.19 所示。

图 1.17　HTTPS 规则

图 1.18　SSH 规则

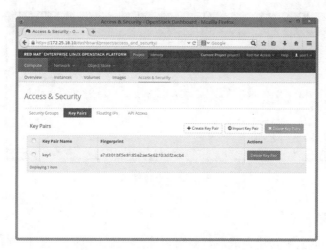

图 1.19　创建密钥对

（3）配置浮动 IP 地址

在 Floating IPs 选项卡中，单击 Allocate IP To Project 来定义实例的浮动 IP 地址，可获取从 public 网段定义外网 IP 地址范围中的地址，可以设置多个浮动 IP 地址。这里获得指定网段范围 172.25.18.25～172.25.18.99 中的第二个 IP 地址 172.25.18.26，这个地址将成为外网用户访问虚拟机实例的 IP 地址，如图 1.20 所示。

图 1.20　浮动 IP 地址

3．上传镜像

OpenStack 使用 Glance 服务管理镜像创建以及重构实例，为了使用方便允许用户上传一定数量的镜像文件来提供创建实例使用。在 Project 选项卡中的 Compute 子选项卡选择 images，通过 Create Image 来上传镜像文件，如图 1.21 所示。

图 1.21　上传镜像

4．创建实例

将上述的所有资源创建完成后，就可以实现实例的创建，即将所有资源整合分配给实例使用，创建实例参数如表 1-9 所示。

表 1-9　创建实例参数

参数	值
实例映像	Image1
实例名称	Small
实例类别	M1.small
实例密钥对	Key1
实例安全组	Sec1
实例浮动 IP 地址	172.25.18.26
卷名称	Myvol1
卷大小	2GB
卷快照名称	Myvol1-snap1

首先需要分别设置 servera 与 serverb 主机支持 KVM 虚拟化功能。

```
[root@servera ~]# ll /dev/kvm
crw-rw-rw-. 1 root kvm 10, 232 Feb 18 13:13 /dev/kvm
[root@servera ~]# crudini --set /etc/nova/nova.conf libvirt virt_type kvm
[root@servera ~]# openstack-service restart nova

[root@serverb ~]# ll /dev/kvm
crw-rw-rw-. 1 root kvm 10, 232 Feb 18 13:12 /dev/kvm
[root@serverb ~]# crudini --set /etc/nova/nova.conf libvirt virt_type kvm
[root@serverb ~]# openstack-service restart nova
```

然后在 Project 选项卡下的 Compute 子选项卡下的 Instance 中，单击 Launch Instance 来创建实例，需要分别指定实例的细则（Details）即实例名称、规格、创建实例的数量，运行实例的镜像，如图 1.22 所示。

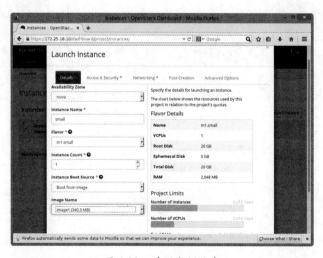

图 1.22　填写实例信息

指定实例所运行的安全组（Access & Security），即确定该实例防火墙所开放的端口，如图 1.23 所示。

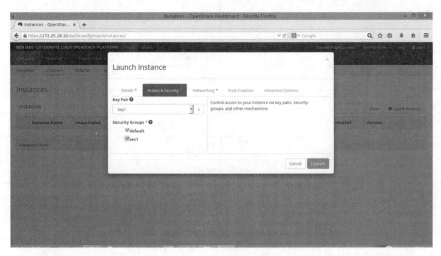

图 1.23　勾选安全组

指定实例运行网络（Networking），实例运行在网络内部，所以添加实例网卡为 private 网段，如图 1.24 所示。

图 1.24　选择网络

将所有指定资源配置完毕后，单击 Launch 按钮加载启动实例，并且单击 Floating IP 关联此实例的浮动 IP 地址，如图 1.25 所示。

待实例创建完成，可以从图 1.26 所示页面，看到实例的运行状态为 Running，此实例的内网地址为 192.168.0.2，关联的浮动 IP 地址为 172.5.18.26。

此时查看 Network Topology，可以看到完整的网络拓扑，实例运行在内网可通过虚拟路由器与外网相互连通，如图 1.27 所示。

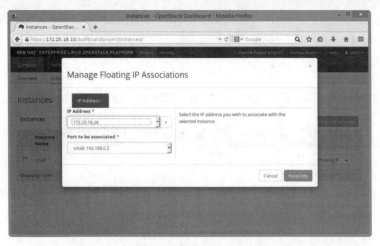

图 1.25　关联浮动 IP 地址

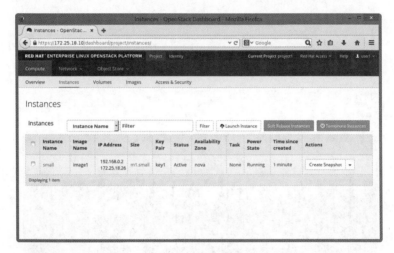

图 1.26　已启动的实例

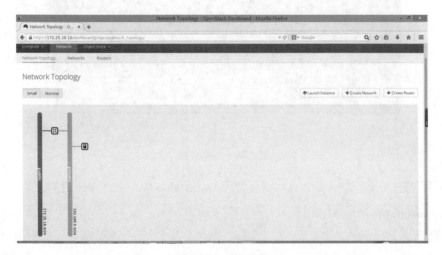

图 1.27　网络拓扑

可以使用 Web 和字符的形式访问实例，Web 访问需要单击实例名称，从"Instance Details：实例名"页面的 Console 口连接实例。添加实例之后以 VNC 协议连接到实例，如图 1.28 所示。输入用户名密码即可登录实例使用。

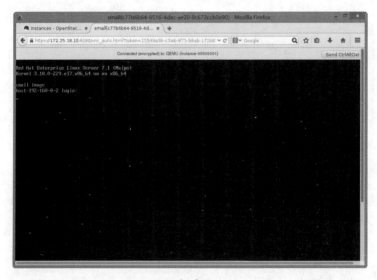

（一）

（二）

图 1.28　连接实例

使用字符的形式访问实例的，需要使用 workstaion 中下载的 key 连接实例。

[root@workstation Downloads]# ls
key1.pem

为了确保 key 的安全性，这里更改 key 的权限为所有者可读写，其他所有用户均无任何权限。

[root@workstation Downloads]# chmod 600 key1.pem

使用 SSH 协议连接时，-i 选项来指定使用密钥。

[root@workstation Downloads]# ssh -i key1.pem root@172.25.18.26
The authenticity of host '172.25.18.26 (172.25.18.26)' can't be established.
ECDSA key fingerprint is 44:bb:59:5b:53:3e:f3:e5:aa:e1:ce:58:9e:f9:b3:ca.
Are you sure you want to continue connecting (yes/no)? yes
Warning: Permanently added '172.25.18.26' (ECDSA) to the list of known hosts.
Please login as the user "cloud-user" rather than the user "root".
Connection to 172.25.18.26 closed.

提示使用的登录用户为 cloud-user。

[root@workstation Downloads]# ssh -i key1.pem cloud-user@172.25.18.26
[cloud-user@small ~]$

1.2.3　扩展应用

需求：给实例添加硬盘设备，并且创建快照。上传带 Web 服务的镜像运行实例。

在 Project 选项卡下的 Compute 子选项卡中，选择 Volumes，单击 Create Volume 按钮，填写好添加硬盘的名称以及大小，就可以创建一块固定大小的硬盘设备，如图 1.29 所示。

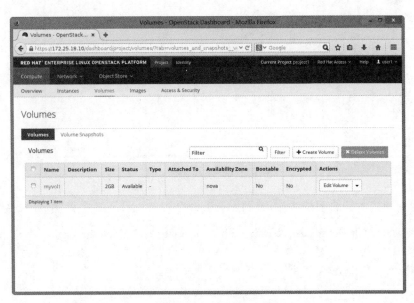

图 1.29　创建硬盘

给硬盘创建快照的操作需要在硬盘附加在实例之前进行，在新添加卷的 Actions 下拉列表中选择 Create Snapshot，填写快照名称即可。创建完成后，在 Volume Snapshots 中可以查看快照信息，如图 1.30 所示。

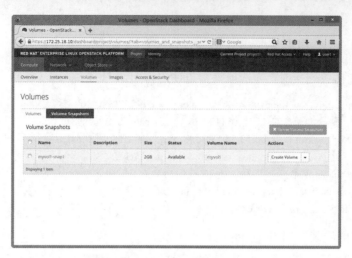

图 1.30　创建快照

在新添加卷的 Actions 下拉列表中选择 Attach to Instance，将新建的卷附加到实例上，附加实例后的信息如图 1.31 所示。

图 1.31　附加硬盘到实例

返回到连接实例的字符终端，运行 lsblk 命令可以查看连接至该实例的磁盘信息。

```
[root@workstation Downloads]# ssh -i key1.pem cloud-user@172.25.18.26
[cloud-user@small ~]$ lsblk
NAME   MAJ:MIN RM  SIZE RO TYPE MOUNTPOINT
vda    253:0    0   20G  0 disk
?..vda1 253:1   0   20G  0 part /
vdb    253:16   0  512M  0 disk [SWAP]
vdc    253:32   0    2G  0 disk
```

以相同方式上传一个带有 Web 环境的实例映像 web.img 并创建实例，如图 1.32 所示。

图 1.32　运行 Web 实例

使用字符终端连接该实例，可以查看该实例上运行的 Web 进程。

[root@workstation Downloads]# ssh -i key1.pem cloud-user@172.25.18.27
The authenticity of host '172.25.18.27 (172.25.18.27)' can't be established.
ECDSA key fingerprint is 1d:72:f5:ee:d3:18:5b:ec:74:d2:8f:0e:85:3b:08:80.
Are you sure you want to continue connecting (yes/no)? yes
Warning: Permanently added '172.25.18.27' (ECDSA) to the list of known hosts.
[cloud-user@web ~]$ su - root
Password:
[root@web ~]# netstat -antpu |grep httpd
tcp6   0   0 :::80   :::*   LISTEN   720/httpd
tcp6   0   0 :::443   :::*   LISTEN   720/httpd
[root@web ~]#

该 Web 服务可以在 workstation 主机通过浏览器形式进行访问。如图 1.33 所示。

[root@workstation ~]# firefox https://172.25.18.27

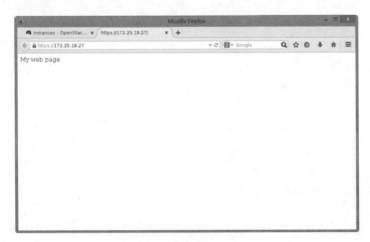

图 1.33　验证 Web 服务

本章总结

- 云计算模型中有三种基本服务模型：IaaS、PaaS、SaaS。
- OpenStack 主要包含计算服务、网络服务、身份认证服务、控制面板服务、镜像服务、块存储服务、对象存储服务和计量服务，每个服务都有一个完整的项目支撑。
- 红帽可以使用 packstack 工具自动安装部署 OpenStack 环境。

本章作业

在 workstaion 上 ping 172.25.18.26 可以通吗？如何可以 ping 通？

第 2 章

OpenStack 搭建企业私有云

技能目标
- 理解 OpenStack 各组件的功能
- 能够手工部署 OpenStack 各组件

本章导读

　　上一章我们体验了 OpenStack 的自动安装，本章将介绍手工部署 OpenStack 各组件，使大家在部署过程中深刻理解 OpenStack 各组件的功能。

知识服务

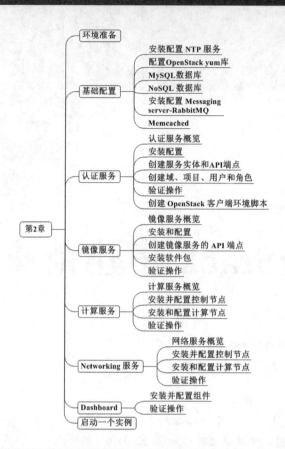

2.1 环境准备

使用 3 台主机，如表 2-1 所示。

表 2-1 主机配置

主机名	操作系统	IP	备注
controller	CentOS 7.2 x86_64	外网 eno16777728：192.168.200.10/24 内网 eno33554952：10.10.10.10/8	控制节点
compute	CentOS 7.2 x86_64	外网 eno16777728：192.168.200.12/24 内网 eno33554952：10.10.10.12/8	计算节点
compute	CentOS 7.2 x86_64	外网 eno16777728：192.168.200.11/24 内网 eno33554952：10.10.10.11/8	块存储节点

1. 所有主机关闭防火墙与 SELinux

```
[root@controller ~]# systemctl disable firewalld.service
[root@controller ~]# systemctl stop firewalld.service
[root@controller ~]# sed -i '/SELINUX/s/enforcing/disabled/' /etc/selinux/config
[root@controller ~]# setenforce 0
```

2. 控制节点配置网络接口

```
[root@localhost ~]# cat /etc/sysconfig/network-scripts/ifcfg-eno16777728
TYPE="Ethernet"
BOOTPROTO="static"
DEFROUTE="yes"
PEERDNS="yes"
PEERROUTES="yes"
IPV4_FAILURE_FATAL="no"
IPV6INIT="no"
IPV6_AUTOCONF="no"
IPV6_DEFROUTE="no"
IPV6_PEERDNS="no"
IPV6_PEERROUTES="no"
IPV6_FAILURE_FATAL="no"
NAME="eno16777728"
UUID="ee81d0c8-c436-4bec-8a4a-aefbb109a94a"
DEVICE="eno16777728"
ONBOOT="yes"
IPADDR=192.168.200.10
NETMASK=255.255.255.0
GATEWAY=192.168.1.1
DNS1=202.106.0.20

[root@localhost ~]# cat /etc/sysconfig/network-scripts/ifcfg-eno33554952
TYPE=Ethernet
BOOTPROTO=static
DEFROUTE=yes
PEERDNS=yes
PEERROUTES=yes
IPV4_FAILURE_FATAL=no
IPV6INIT=no
IPV6_AUTOCONF=no
IPV6_DEFROUTE=no
IPV6_PEERDNS=no
IPV6_PEERROUTES=no
IPV6_FAILURE_FATAL=no
NAME=eno33554952
UUID=8deb1b8d-f7cc-4d1c-b376-5647a814faa1
DEVICE=eno33554952
ONBOOT=yes
IPADDR=10.10.10.10
MASK=255.0.0.0
[root@localhost ~]#systemctl restart network.service
```

3. 计算节点配置网络接口

[root@localhost ~]# cat /etc/sysconfig/network-scripts/ifcfg-eno16777728
TYPE="Ethernet"
BOOTPROTO="static"
DEFROUTE="yes"
PEERDNS="yes"
PEERROUTES="yes"
IPV4_FAILURE_FATAL="no"
IPV6INIT="no"
IPV6_AUTOCONF="no"
IPV6_DEFROUTE="no"
IPV6_PEERDNS="no"
IPV6_PEERROUTES="no"
IPV6_FAILURE_FATAL="no"
NAME="eno16777728"
UUID="948305fc-ed49-48c7-943e-18974296e544"
DEVICE="eno16777728"
ONBOOT="yes"
IPADDR=192.168.200.12
NETMASK=255.255.255.0
GATEWAY=192.168.1.1
DNS1=202.106.0.20

[root@localhost ~]# cat /etc/sysconfig/network-scripts/ifcfg-eno33554952
TYPE=Ethernet
BOOTPROTO=static
DEFROUTE=yes
PEERDNS=yes
PEERROUTES=yes
IPV4_FAILURE_FATAL=no
IPV6INIT=no
IPV6_AUTOCONF=no
IPV6_DEFROUTE=no
IPV6_PEERDNS=no
IPV6_PEERROUTES=no
IPV6_FAILURE_FATAL=no
NAME=eno33554952
UUID=7798854f-a566-4408-b4b6-49ceef36251b
DEVICE=eno33554952
ONBOOT=yes
IPADDR=10.10.10.12
NETMASK=255.255.255.0
[root@localhost ~]#systemctl restart network.service

4. 块存储节点配置网络接口

```
[root@block ~]# cat /etc/sysconfig/network-scripts/ifcfg-eno16777736
TYPE="Ethernet"
BOOTPROTO="static"
DEFROUTE="yes"
PEERDNS="yes"
PEERROUTES="yes"
IPV4_FAILURE_FATAL="no"
IPV6INIT="no"
IPV6_AUTOCONF="no"
IPV6_DEFROUTE="no"
IPV6_PEERDNS="no"
IPV6_PEERROUTES="no"
IPV6_FAILURE_FATAL="no"
NAME="eno16777736"
UUID="532635cd-06e8-4371-a509-db91e5e002d8"
DEVICE="eno16777736"
ONBOOT="yes"
IPADDR=192.168.200.11
NETMASK=255.255.255.0
GATEWAY=192.168.200.1
DNS1=202.106.0.20
[root@block ~]# cat /etc/sysconfig/network-scripts/ifcfg-eno33554944
TYPE=Ethernet
BOOTPROTO=static
DEFROUTE=yes
PEERDNS=yes
PEERROUTES=yes
IPV4_FAILURE_FATAL=no
IPV6INIT=no
IPV6_AUTOCONF=no
IPV6_DEFROUTE=no
IPV6_PEERDNS=no
IPV6_PEERROUTES=no
IPV6_FAILURE_FATAL=no
NAME=eno33554944
UUID=523c34cf-0bbe-447a-8db4-73772470a9ae
DEVICE=eno33554944
ONBOOT=yes
IPADDR=10.10.10.11
NETMASK=255.0.0.0
```

5. 配置域名解析

（1）所有主机修改 /etc/hosts 文件

```
[root@controller ~]# vim /etc/hosts
127.0.0.1    localhost localhost.localdomain localhost4 localhost4.localdomain4
::1          localhost localhost.localdomain localhost6 localhost6.localdomain6
10.10.10.10 controller
10.10.10.11 block
10.10.10.12 compute

192.168.200.10 controller
192.168.200.11 block
192.168.200.12 compute
```

（2）设置节点主机名

修改控制节点：

```
[root@localhost ~]# hostname controller
[root@localhost ~]# bash
[root@controller ~]# cat /etc/hostname
controller
[root@controller ~]# reboot      // 需要重启计算机，否则 RabbitMQ 创建用户报错
```

修改计算节点：

```
[root@localhost ~]# hostname compute
[root@localhost ~]# bash
[root@compute ~]# cat /etc/hostname
compute
[root@compute ~]# reboot
```

2.2 基础配置

2.2.1 安装配置 NTP 服务

安装 Chrony，使不同节点之间实现 NTP 方案。建议配置控制器节点为 NTP 服务器，然后其他节点指定节点实现时间同步。

（1）在 controller 节点安装配置 chrony

```
[root@controller ~]# yum -y install chrony
[root@controller ~]# vim /etc/chrony.conf
server controller iburst      // 指定时间同步主机为本机，控制节点默认跟公共服务器池同步时间
allow 10.10.10.0/24           // 允许其他节点可以连接到控制节点的 chrony 后台进程
```

```
[root@controller ~]# systemctl enable chronyd.service
[root@controller ~]# systemctl start chronyd.service
[root@controller ~]# chronyc sources
210 Number of sources = 4
MS Name/IP address    Stratum  Poll  Reach  Last  Rx            Last sample
===============================================================================
^? 109.205.112.69     0        8     0      10y   +0ns[  +0ns]  +/- 0ns
^- 125.62.193.121     2        6     144    216   +401ms[+113ms] +/- 486ms
^* ntp.tums.ac.ir     2        6     257    85    +3104us[-401ms] +/- 323ms
^? 103.47.76.177      2        7     40     232   +428ms[+28378s] +/- 362ms
```

（2）在 compute 和 block 节点指定 NTP 主机

```
[root@compute ~]# yum install chrony –y
[root@compute ~]# vim /etc/chrony.conf
server controller iburst
[root@compute ~]# systemctl enable chronyd.service
[root@compute ~]# systemctl start chronyd.service
[root@compute ~]# chronyc sources
210 Number of sources = 5
MS Name/IP address    Stratum  Poll  Reach  Last  Rx            Last sample
===============================================================================
^? 103.47.76.177      0        6     0      10y   +0ns[  +0ns] +/- 0ns
^? ntp.nic.kz         0        6     0      10y   +0ns[  +0ns] +/- 0ns
^? 185.23.153.237     0        6     0      10y   +0ns[  +0ns] +/- 0ns
^? 106.247.248.106    3        6     3      1     -88ms[ -88ms] +/- 243ms
^? controller         0        6     0      10y   +0ns[  +0ns] +/- 0ns
```

在 Name/IP address 列的内容应显示 NTP 服务器的主机名或者 IP 地址。在 S 列的内容应该在 NTP 服务目前同步的上游服务器前显示 "*"。

2.2.2 配置 OpenStack yum 库

在 CentOS 中，extras 仓库提供用于启用 OpenStack 仓库的 RPM 包。CentOS 默认启用 extras 仓库，因此可以直接安装用于启用 OpenStack 仓库的包。

```
[root@controller ~]#cd /etc/yum.repos.d/
[root@controller yum.repos.d]# mkdir bak
[root@controller yum.repos.d]# mv *.repo bak
[root@controller ~]# wget -O /etc/yum.repos.d/CentOS-Base.repo http://mirrors.aliyun.com/repo/
            Centos-7.repo
[root@controller ~]# yum clean all && yum makecache
[root@controller ~]# yum -y install centos-release-openstack-mitaka
```

在主机上升级包：

```
[root@controller ~]# yum -y upgrade
```

软件包镜像站找不到解决方案：

```
[root@controller ~]# cat tools.sh
#!/bin/bash
url="http://mirrors.aliyun.com/centos/7.2.1511/cloud/x86_64/openstack-liberty/common/"
for tools in $(awk '{print $1}' error | sed 's/:/.rpm/g')
do
    yum -y install $url/$tools
done
```

需把报错软件包放置于 error 文件中：

```
[root@controller ~]# cat error
  python-ncclient-0.4.2-2.el7.noarch: [Errno 256] No more mirrors to try.
  python2-pysaml2-3.0.2-2.el7.noarch: [Errno 256] No more mirrors to try.
  erlang-asn1-18:3.4.4-1.el7.x86_64: [Errno 256] No more mirrors to try.
```

安装 OpenStack 客户端：

```
[root@controller ~]# yum -y install python-openstackclient
```

RHEL 和 CentOS 默认启用了 SELinux，安装 openstack-selinux 软件包以便自动管理 OpenStack 服务的安全策略：

```
[root@controller ~]# yum -y install openstack-selinux
```

2.2.3 MySQL 数据库

大多数 OpenStack 服务使用 SQL 数据库来存储信息。数据库需运行在控制节点上。

（1）控制器节点安装 database：mariadb

```
[root@controller ~]# yum install mariadb mariadb-server python2-PyMySQL -y
```

（2）创建并编辑 /etc/my.cnf.d/openstack.cnf，然后完成如下操作

在 [mysqld] 部分，设置 bind-address 值为控制节点的管理网络 IP 地址以使得其他节点可以通过管理网络访问数据库：

```
[root@controller ~]# vim /etc/my.cnf.d/openstack.cnf
[mysqld]
bind-address = 10.10.10.10                  // 设置监听地址
default-storage-engine = innodb             // 设置默认存储引擎
innodb_file_per_table                       // 设置独享表空间
max_connections = 4096                      // 设置最大连接数
collation-server = utf8_general_ci          // 设置校对规则
character-set-server = utf8                 // 设置创建数据库时的默认字符集
```

（3）重启 mariadb 服务并设置开机启动

```
[root@controller ~]# systemctl enable mariadb.service
[root@controller ~]# systemctl start mariadb.service
```

```
[root@controller ~]# systemctl status mariadb.service
  mariadb.service - MariaDB 10.1 database server
   Loaded: loaded (/usr/lib/systemd/system/mariadb.service; enabled; vendor preset: disabled)
   Active: active (running）since 一 2016-11-28 21:34:07 CST; 2s ago
  Process: 2117 ExecStartPost=/usr/libexec/mysql-check-upgrade (code=exited, status=0/SUCCESS)
  Process: 1922 ExecStartPre=/usr/libexec/mysql-prepare-db-dir %n (code=exited, status=0/SUCCESS)
  Process: 1900 ExecStartPre=/usr/libexec/mysql-check-socket (code=exited, status=0/SUCCESS)
 Main PID: 2089 (mysqld)
   Status: "Taking your SQL requests now..."
   CGroup: /system.slice/mariadb.service
           └─2089 /usr/libexec/mysqld --basedir=/usr

11 月 28 21:34:06 controller mysql-prepare-db-dir[1922]: Corporation Ab. You can contact us about
   this at sales@mariadb.com.
11 月 28 21:34:06 controller mysql-prepare-db-dir[1922]: Alternatively consider joining our
   community based development effort:
11 月 28 21:34:06 controller mysql-prepare-db-dir[1922]: http://mariadb.com/kb/en/contributing-to-
   the-mariadb-project/
11 月 28 21:34:06 controller mysqld[2089]: error: Found option without preceding group in config file:
   /etc/my.cnf.d/op...ine: 1
11 月 28 21:34:06 controller mysqld[2089]: 2016-11-28 21:34:06 140061270055040 [Note] /usr/
   libexec/mysqld (mysqld 10.1....89 ...
11 月 28 21:34:06 controller mysql-check-upgrade[2117]: error: Found option without preceding group
   in config file: /etc...ne: 1
11 月 28 21:34:06 controller mysql-check-upgrade[2117]: error: Found option without preceding group
   in config file: /etc...ne: 1
11 月 28 21:34:06 controller mysql-check-upgrade[2117]: error: Found option without preceding group
   in config file: /etc...ne: 1
11 月 28 21:34:06 controller mysql-check-upgrade[2117]: error: Found option without preceding group
   in config file: /etc...ne: 1
11 月 28 21:34:07 controller systemd[1]: Started MariaDB 10.1 database server.
Hint: Some lines were ellipsized, use -l to show in full.
```

（4）对数据库进行安全加固

为了保证数据库服务的安全性，运行 mysql_secure_installation 脚本。特别需要说明的是，为数据库的 root 用户设置一个适当的密码。

```
[root@controller ~]# mysql_secure_installation
error: Found option without preceding group in config file: /etc/my.cnf.d/openstack.cnf at line: 1

NOTE: RUNNING ALL PARTS OF THIS SCRIPT IS RECOMMENDED FOR ALL MariaDB
      SERVERS IN PRODUCTION USE!  PLEASE READ EACH STEP CAREFULLY!

In order to log into MariaDB to secure it, we'll need the current
password for the root user.  If you've just installed MariaDB, and
you haven't set the root password yet, the password will be blank,
so you should just press enter here.
```

```
Enter current password for root (enter for none):
OK, successfully used password, moving on...

Setting the root password ensures that nobody can log into the MariaDB
root user without the proper authorisation.

Set root password? [Y/n] y
New password:                          //123456
Re-enter new password:                 //123456
Password updated successfully!
Reloading privilege tables..
 ... Success!


By default, a MariaDB installation has an anonymous user, allowing anyone
to log into MariaDB without having to have a user account created for
them.  This is intended only for testing, and to make the installation
go a bit smoother.  You should remove them before moving into a
production environment.

Remove anonymous users? [Y/n] y
 ... Success!

Normally, root should only be allowed to connect from 'localhost'.  This
ensures that someone cannot guess at the root password from the network.

Disallow root login remotely? [Y/n] y
 ... Success!

By default, MariaDB comes with a database named 'test' that anyone can
access.  This is also intended only for testing, and should be removed
before moving into a production environment.

Remove test database and access to it? [Y/n] y
 - Dropping test database...
 ... Success!
 - Removing privileges on test database...
 ... Success!

Reloading the privilege tables will ensure that all changes made so far
will take effect immediately.

Reload privilege tables now? [Y/n] y
 ... Success!
```

```
Cleaning up...

All done! If you've completed all of the above steps, your MariaDB
installation should now be secure.

Thanks for using MariaDB!
```

2.2.4 NoSQL 数据库

Telemetry 服务使用 NoSQL 数据库来存储信息，通常这个数据库运行在控制节点上。向导中使用 MongoDB。只有安装 Telemetry 服务时，才需要安装 NoSQL 数据库服务。

（1）安装 MongoDB 包

```
[root@localhost ~]# yum install mongodb-server mongodb -y
```

（2）编辑文件 /etc/mongod.conf 并完成如下操作

配置 bind_ip 使用控制节点管理网卡的 IP 地址。

```
6 bind_ip = 10.10.10.10
```

默认情况下，MongoDB 会在 /var/lib/mongodb/journal 目录下创建几个 1GB 大小的日志文件。如果想将每个日志文件大小减小到 128MB 并且限制日志文件占用的总空间为 512MB，可以配置 smallfiles 的值：

```
113 smallfiles = true
```

（3）完成安装

启动 MongoDB 并配置它随系统启动：

```
[root@controller ~]# systemctl enable mongod.service
[root@controller ~]# systemctl start mongod.service
```

2.2.5 安装配置 Messaging server-RabbitMQ

OpenStack 使用 message queue 协调操作和各服务的状态信息。消息队列服务一般运行在控制节点上。OpenStack 支持好几种消息队列服务包括 RabbitMQ、Qpid 和 ZeroMQ。不过，大多数发行版本的 OpenStack 包支持特定的消息队列服务。安装 RabbitMQ 消息队列服务，因为大部分发行版本都支持它。

（1）在 controller 端安装 RabbitMQ

```
[root@controller ~]# yum -y install rabbitmq-server
```

（2）启动消息队列服务并将其配置为随系统启动

```
[root@controller ~]# systemctl enable rabbitmq-server.service
[root@controller ~]# systemctl start rabbitmq-server.service
```

```
[root@controller ~]# systemctl status rabbitmq-server.service
  rabbitmq-server.service - RabbitMQ broker
  Loaded: loaded (/usr/lib/systemd/system/rabbitmq-server.service; enabled; vendor preset: disabled)
  Active: active (running) since 一 2016-11-28 21:47:39 CST; 1s ago
 Main PID: 2329 (beam)
  Status: "Initialized"
  CGroup: /system.slice/rabbitmq-server.service
       ├─2329 /usr/lib64/erlang/erts-7.3.1.2/bin/beam -W w -A 64 -P 1048576 -t 5000000 -stbt db -K true -- -root /usr/lib...
       ├─2511 inet_gethost 4
       └─2512 inet_gethost 4

11 月 28 21:47:26 controller systemd[1]: Starting RabbitMQ broker...
11 月 28 21:47:33 controller rabbitmq-server[2329]: RabbitMQ 3.6.5. Copyright (C) 2007-2016 Pivotal Software, Inc.
11 月 28 21:47:33 controller rabbitmq-server[2329]: ## ##      Licensed under the MPL.  See http://www.rabbitmq.com/
11 月 28 21:47:33 controller rabbitmq-server[2329]: ## ##
11 月 28 21:47:33 controller rabbitmq-server[2329]: ##########  Logs: /var/log/rabbitmq/rabbit@controller.log
11 月 28 21:47:33 controller rabbitmq-server[2329]: ######  ##        /var/log/rabbitmq/rabbit@controller-sasl.log
11 月 28 21:47:33 controller rabbitmq-server[2329]: ##########
11 月 28 21:47:33 controller rabbitmq-server[2329]: Starting broker...
11 月 28 21:47:39 controller systemd[1]: Started RabbitMQ broker.
11 月 28 21:47:39 controller rabbitmq-server[2329]: completed with 0 plugins.
```

（3）RabbitMQ 添加 openstack 用户，密码 123456

如遇到创建用户错误，需修改永久主机名重启当前主机。

```
[root@controller ~]# rabbitmqctl add_user openstack 123456
Error: unable to connect to node rabbit@localhost: nodedown

DIAGNOSTICS
===========

attempted to contact: [rabbit@localhost]

rabbit@localhost:
  * connected to epmd (port 4369) on localhost
  * epmd reports node 'rabbit' running on port 25672
  * TCP connection succeeded but Erlang distribution failed
  * suggestion: hostname mismatch?
  * suggestion: is the cookie set correctly?
  * suggestion: is the Erlang distribution using TLS?
```

```
current node details:
- node name: 'rabbitmq-cli-14@controller'
- home dir: /var/lib/rabbitmq
- cookie hash: eLzpCVvfgDq8buH/1uoI5w==


[root@controller ~]# rabbitmqctl add_user openstack 123456
Creating user "openstack" ...
```

（4）给 openstack 用户配置写和读权限

```
[root@controller ~]# rabbitmqctl set_permissions openstack ".*" ".*" ".*"
Setting permissions for user "openstack" in vhost "/" ...
```

2.2.6 memcached

缓存服务 memcached 运行在控制节点。在生产部署中，我们推荐联合启用防火墙、认证和加密保证它的安全。

（1）安装软件包

```
[root@controller ~]# yum install memcached python-memcached
```

（2）启动 memcached 服务，并且配置它随机启动

```
[root@controller ~]# systemctl enable memcached.service
[root@controller ~]# systemctl start memcached.service
[root@controller ~]# systemctl status memcached.service
● memcached.service - Memcached
   Loaded: loaded (/usr/lib/systemd/system/memcached.service; enabled; vendor preset: disabled)
   Active: active (running) since 一 2016-11-28 22:01:46 CST; 8s ago
 Main PID: 2479 (memcached)
   CGroup: /system.slice/memcached.service
           └─2479 /usr/bin/memcached -u memcached -p 11211 -m 64 -c 1024

11 月 28 22:01:46 controller systemd[1]: Started Memcached.
11 月 28 22:01:46 controller systemd[1]: Starting Memcached...
```

2.3 认证服务

2.3.1 认证服务概览

Identity 服务为其他 OpenStack 服务提供验证和授权服务，为所有服务提供终端目录。其他 OpenStack 服务将身份认证服务当做通用统一 API 来使用。此外，提供用户信息但是不在 OpenStack 项目中的服务（如 LDAP 服务）可被整合进先前存在的基础

设施中。

为了从 Identity 服务中获益，其他的 OpenStack 服务需要与它合作。当某个 OpenStack 服务收到来自用户的请求时，该服务询问 Identity 服务，验证该用户是否有权限进行此次请求，身份服务包含这些组件：

- **服务器**：一个中心化的服务器使用 RESTful 接口来提供认证和授权服务。
- **驱动**：驱动或服务后端被整合进集中式服务器中。它们被用来访问 OpenStack 外部仓库的身份信息，并且它们可能已经存在于 OpenStack 被部署的基础设施（例如，SQL 数据库或 LDAP 服务器）中。
- **模块**：中间件模块运行于使用身份认证服务的 OpenStack 组件的地址空间中。这些模块拦截服务请求，取出用户凭据，并将它们送入中央服务器寻求授权。中间件模块和 OpenStack 组件间的整合使用 Python Web 服务器网关接口。

当安装 OpenStack 身份服务时，用户必须将之注册到其 OpenStack 安装环境的每个服务。身份服务才可以追踪哪些 OpenStack 服务已经安装，以及在网络中定位它们。

出于性能原因，这个配置部署 Fernet 令牌和 Apache HTTP 服务处理请求。

2.3.2 安装配置

1. 配置 MySQL 数据库及授权

在配置 OpenStack 身份认证服务前，必须创建一个数据库和管理员令牌。

以 root 用户连接到数据库服务器：

```
[root@controller ~]# mysql -u root -p123456
```

创建 keystone 数据库：

```
MariaDB [(none)]> CREATE DATABASE keystone;
Query OK, 1 row affected (0.00 sec)
```

对 keystone 数据库授予权限：

```
MariaDB [(none)]> GRANT ALL PRIVILEGES ON keystone.* TO 'keystone'@'localhost'
    IDENTIFIED BY '123456';
Query OK, 0 rows affected (0.01 sec)
MariaDB [(none)]> GRANT ALL PRIVILEGES ON keystone.* TO 'keystone'@'%' IDENTIFIED BY
    '123456';
Query OK, 0 rows affected (0.00 sec)
```

2. 生成随机值作为管理员的令牌

```
[root@controller ~]# openssl rand -hex 10
294a4c8a8a475f9b9836
```

3. 安装软件包

```
[root@controller ~]# yum install openstack-keystone httpd mod_wsgi
```

4. 编辑文件 /etc/keystone/keystone.conf 并完成如下操作

```
[root@controller ~]# cp /etc/keystone/keystone.conf{,.bak}
[root@controller ~]# vim /etc/keystone/keystone.conf
[DEFAULT]
  13 admin_token = 294a4c8a8a475f9b9836           // 定义初始管理令牌的值

[database]
 549 connection = mysql://keystone:123456@controller/keystone   // 配置数据库访问

[token]
2005 provider = fernet                            // 配置 Fernet UUID 令牌的提供者
```

5. 初始化身份认证服务的数据库

```
[root@controller ~]# su -s /bin/sh -c "keystone-manage db_sync" keystone
```

6. 初始化 Fernet keys

```
[root@controller ~]# keystone-manage fernet_setup --keystone-user keystone --keystone-group keystone
```

7. 配置 Apache HTTP server

（1）编辑配置文件 /etc/httpd/conf/httpd.conf

```
[root@controller ~]# vim /etc/httpd/conf/httpd.conf
ServerName controller
```

（2）编辑配置文件 /etc/httpd/conf.d/wsgi-keystone.conf

```
[root@controller ~]# cat /etc/httpd/conf.d/wsgi-keystone.conf
Listen 5000
Listen 35357

<VirtualHost *:5000>
    WSGIDaemonProcess keystone-public processes=5 threads=1 user=keystone group=keystone
        display-name=%{GROUP}
    WSGIProcessGroup keystone-public
    WSGIScriptAlias / /usr/bin/keystone-wsgi-public
    WSGIApplicationGroup %{GLOBAL}
    WSGIPassAuthorization On
    ErrorLogFormat "%{cu}t %M"
    ErrorLog /var/log/httpd/keystone-error.log
    CustomLog /var/log/httpd/keystone-access.log combined

    <Directory /usr/bin>
        Require all granted
    </Directory>
```

```
    </VirtualHost>

    <VirtualHost *:35357>
      WSGIDaemonProcess keystone-admin processes=5 threads=1 user=keystone group=keystone
          display-name=%{GROUP}
      WSGIProcessGroup keystone-admin
      WSGIScriptAlias / /usr/bin/keystone-wsgi-admin
      WSGIApplicationGroup %{GLOBAL}
      WSGIPassAuthorization On
      ErrorLogFormat "%{cu}t %M"
      ErrorLog /var/log/httpd/keystone-error.log
      CustomLog /var/log/httpd/keystone-access.log combined

    <Directory /usr/bin>
        Require all granted
    </Directory>
    </VirtualHost>
```

(3) 重启 apache 服务

```
[root@controller ~]# systemctl enable httpd.service
[root@controller ~]# systemctl start httpd.service
```

2.3.3 创建服务实体和 API 端点

身份认证服务提供服务的目录和它们的位置。添加到 OpenStack 环境中的服务在目录中需要一个 service 实体和一些 API 端点。

配置认证令牌：

```
[root@controller ~]# export OS_TOKEN=294a4c8a8a475f9b9836
```

配置端点 URL：

```
[root@controller ~]# export OS_URL=http://controller:35357/v3
```

配置认证 API 版本：

```
[root@controller ~]# export OS_IDENTITY_API_VERSION=3
```

（1）在 OpenStack 环境中，认证服务管理服务目录。服务使用这个目录来决定环境中可用的服务。

创建服务实体和身份认证服务：

```
[root@controller ~]# openstack service create   --name keystone --description "OpenStack Identity"
   identity
+-------------+----------------------------------+
| Field       | Value                            |
+-------------+----------------------------------+
| description | OpenStack Identity               |
```

```
| enabled  | True                             |
| id       | 25763c02ad2b496a8916acd651a679dc |
| name     | keystone                         |
| type     | identity                         |
+----------+----------------------------------+
```

OpenStack 是动态生成 ID 的，因此实际操作看到的输出会与示例中的命令行输出不相同。

（2）身份认证服务管理了一个与实际操作环境相关的 API 端点的目录。服务使用这个目录来决定如何与实际操作环境中的其他服务进行通信。

OpenStack 使用三个 API 端点变种代表每种服务：admin、internal 和 public。默认情况下，管理 API 端点允许修改用户和租户而公共和内部 APIs 不允许这些操作。在生产环境中，出于安全原因，变种为了服务不同类型的用户可能驻留在单独的网络上。对实例而言，公共 API 网络为了让顾客管理他们自己的云，所以在互联网上是可见的。管理 API 网络在管理云基础设施的组织中操作也是有所限制的。内部 API 网络可能会被限制在包含 OpenStack 服务的主机上。此外，OpenStack 支持可伸缩的多区域。为了简单起见，本书中所有端点变种和默认 RegionOne 区域都使用管理网络。

创建认证服务的 API 端点：

```
[root@controller ~]# openstack endpoint create --region RegionOne  identity public
    http://controller:5000/v3
+--------------+----------------------------------+
| Field        | Value                            |
+--------------+----------------------------------+
| enabled      | True                             |
| id           | e263c6a46e3845c28009bf463828af38 |
| interface    | public                           |
| region       | RegionOne                        |
| region_id    | RegionOne                        |
| service_id   | 25763c02ad2b496a8916acd651a679dc |
| service_name | keystone                         |
| service_type | identity                         |
| url          | http://controller:5000/v3        |
+--------------+----------------------------------+

[root@controller ~]# openstack endpoint create --region RegionOne  identity internal
      http://controller:5000/v3
+--------------+----------------------------------+
| Field        | Value                            |
+--------------+----------------------------------+
| enabled      | True                             |
| id           | d755808a560440689c2acc6915e34a9f |
| interface    | internal                         |
| region       | RegionOne                        |
```

```
| region_id   | RegionOne                        |
| service_id  | 25763c02ad2b496a8916acd651a679dc |
| service_name| keystone                         |
| service_type| identity                         |
| url         | http://controller:5000/v3        |
+-------------+----------------------------------+
```

```
[root@controller ~]# openstack endpoint create --region RegionOne   identity admin
        http://controller:35357/v3
+-------------+----------------------------------+
| Field       | Value                            |
+-------------+----------------------------------+
| enabled     | True                             |
| id          | 09d3dd4902754c368f905645aefdff70 |
| interface   | admin                            |
| region      | RegionOne                        |
| region_id   | RegionOne                        |
| service_id  | 25763c02ad2b496a8916acd651a679dc |
| service_name| keystone                         |
| service_type| identity                         |
| url         | http://controller:35357/v3       |
+-------------+----------------------------------+
```

每个添加到 OpenStack 环境中的服务要求一个或多个服务实体和三个认证服务中的 API 端点变种。

2.3.4　创建域、项目、用户和角色

身份认证服务为每个 OpenStack 服务提供认证服务。认证服务使用 domains、projects (tenants)、users 和 roles 的组合。

（1）创建域 default。

```
[root@controller ~]# openstack domain create --description "Default Domain" default
+-------------+----------------------------------+
| Field       | Value                            |
+-------------+----------------------------------+
| description | Default Domain                   |
| enabled     | True                             |
| id          | 7174f90333fd46fd8a5f1ca8989df867 |
| name        | default                          |
+-------------+----------------------------------+
```

（2）在环境中，为进行管理操作，创建管理的项目、用户和角色。

1）创建 admin 项目：

```
[root@controller ~]# openstack project create --domain default --description "Admin Project" admin
+-------------+----------------------------------+
```

```
| Field       | Value                            |
+-------------+----------------------------------+
| description | Admin Project                    |
| domain_id   | 7174f90333fd46fd8a5f1ca8989df867 |
| enabled     | True                             |
| id          | 524aa64844814bc5bf6f73fd9e416410 |
| is_domain   | False                            |
| name        | admin                            |
| parent_id   | 7174f90333fd46fd8a5f1ca8989df867 |
+-------------+----------------------------------+
```

2）创建 admin 用户：

```
[root@controller ~]# openstack user create --domain default --password-prompt admin
User Password:                    // 密码 123456
Repeat User Password:             // 密码 123456
+-----------+----------------------------------+
| Field     | Value                            |
+-----------+----------------------------------+
| domain_id | 7174f90333fd46fd8a5f1ca8989df867 |
| enabled   | True                             |
| id        | c7552cc0ee694bc7b5a855cbbd4165a5 |
| name      | admin                            |
+-----------+----------------------------------+
```

3）创建 admin 角色：

```
[root@controller ~]# openstack role create admin
+-----------+----------------------------------+
| Field     | Value                            |
+-----------+----------------------------------+
| domain_id | None                             |
| id        | 751faf47255543119101c4158dbb7314 |
| name      | admin                            |
+-----------+----------------------------------+
```

4）添加 admin 角色到 admin 项目和用户上：

```
[root@controller ~]# openstack role add --project admin --user admin admin
```

创建的任何角色必须映射到每个 OpenStack 服务配置文件目录下的 policy.json 文件中。默认策略是给予 "admin" 角色大部分服务的管理访问权限。

（3）环境中每个服务包含独有用户的 service 项目。

创建 service 项目：

```
[root@controller ~]# openstack project create --domain default --description "Service Project" service
+-------------+----------------------------------+
| Field       | Value                            |
+-------------+----------------------------------+
| description | Service Project                  |
```

```
| domain_id | 7174f90333fd46fd8a5f1ca8989df867 |
| enabled   | True                              |
| id        | 35d50dc5ba844363a801bc1b54b6d2af |
| is_domain | False                             |
| name      | service                           |
| parent_id | 7174f90333fd46fd8a5f1ca8989df867 |
+-----------+-----------------------------------+
```

（4）常规（非管理）任务应该使用无特权的项目和用户。作为例子，本书创建 demo 项目和用户。

1）创建 demo 项目：

```
[root@controller ~]# openstack project create --domain default   --description "Demo Project" demo
+-------------+----------------------------------+
| Field       | Value                            |
+-------------+----------------------------------+
| description | Demo Project                     |
| domain_id   | 7174f90333fd46fd8a5f1ca8989df867 |
| enabled     | True                             |
| id          | 4aef5906b5504faa9898ebc0fb7317db |
| is_domain   | False                            |
| name        | demo                             |
| parent_id   | 7174f90333fd46fd8a5f1ca8989df867 |
+-------------+----------------------------------+
```

2）创建 demo 用户：

```
[root@controller ~]# openstack user create --domain default --password-prompt demo
User Password:                   // 密码 123456
Repeat User Password:            // 密码 123456
+-----------+----------------------------------+
| Field     | Value                            |
+-----------+----------------------------------+
| domain_id | 7174f90333fd46fd8a5f1ca8989df867 |
| enabled   | True                             |
| id        | 83f3a41a98ce45a995cf7defd0bc4938 |
| name      | demo                             |
+-----------+----------------------------------+
```

3）创建 user 角色：

```
[root@controller ~]# openstack role create user
+-----------+----------------------------------+
| Field     | Value                            |
+-----------+----------------------------------+
| domain_id | None                             |
| id        | 20d6cbb44b8142a1ae1e2c105b146df6 |
| name      | user                             |
+-----------+----------------------------------+
```

4）添加 user 角色到 demo 项目和用户：

[root@controller ~]# openstack role add --project demo --user demo user

2.3.5 验证操作

（1）因为安全性的原因，关闭临时认证令牌机制：

编辑 /etc/keystone/keystone-paste.ini 文件，从 [pipeline:public_api]、[pipeline:admin_api] 和 [pipeline:api_v3] 部分删除 admin_token_auth。

（2）重置 OS_TOKEN 和 OS_URL 环境变量：

[root@controller ~]# unset OS_TOKEN OS_URL

（3）作为 admin 用户，请求认证令牌：

```
[root@controller ~]# openstack --os-auth-url http://controller:35357/v3 --os-project-domain-name
    default --os-user-domain-name default --os-project-name admin --os-username admin token issue
Password:    // 密码 123456
+------------+----------------------------------------------------------------------+
| Field      | Value                                                                |
+------------+----------------------------------------------------------------------+
| expires    | 2016-11-28T17:01:07.802649Z                                          |
| id         | gAAAAABYPFTEPmuJSjRpp7qpWpkRuaVAItj6HIUPAAyBnnVPaQj3Leby0Gk2wq1ew     |
|            | PzvY7e0ETKLBh29lIN5W8HrNq6tFD88cZx3CSn5yGo7xRE                       |
|            | 2OT_YwS_-AdTdfdL6_q940vgJ_B2CKWGgQIH5w9Myxnua0d5BxZlbcVwMQPIig-yUt-  |
|            | pNtO4                                                                |
| project_id | 524aa64844814bc5bf6f73fd9e416410                                     |
| user_id    | c7552cc0ee694bc7b5a855cbbd4165a5                                     |
+------------+----------------------------------------------------------------------+
```

这个命令使用 admin 用户的密码。

（4）作为 demo 用户，请求认证令牌：

```
[root@controller ~]# openstack --os-auth-url http://controller:5000/v3 --os-project-domain-name
    default --os-user-domain-name default --os-project-name demo --os-username demo token issue
Password:    // 密码 123456
+------------+----------------------------------------------------------------------+
| Field      | Value                                                                |
+------------+----------------------------------------------------------------------+
| expires    | 2016-11-28T17:02:56.131903Z                                          |
| id         | gAAAAABYPFUww6WVxmNsD4S-2NBiJkPH_NUekCfzI8-H468XQwyw8XUFSaWN8R        |
|            | MsOIidr6SGZcBM8DHhg_BjV4g-                                           |
|            | VJeu4_R7RtjnLjTaiqm1rYbng4VfbTY3y7TwYJaBW-LTEAL5v_                   |
|            | fMDjLvQXhFlH4tNLP92hzU1PGJQRHEXpTscov9XGKLSQE                        |
| project_id | 4aef5906b5504faa9898ebc0fb7317db                                     |
| user_id    | 83f3a41a98ce45a995cf7defd0bc4938                                     |
+------------+----------------------------------------------------------------------+
```

这个命令使用 demo 用户的密码和 API 端口 5000，这样只会允许对身份认证服务 API 的常规（非管理）访问。

2.3.6 创建 OpenStack 客户端环境脚本

前一节中使用环境变量和命令选项的组合通过 OpenStack 客户端与身份认证服务交互。为了提升客户端操作的效率，OpenStack 支持简单的客户端环境变量脚本即 OpenRC 文件。这些脚本通常包含客户端所有常见的选项，当然也支持独特的选项。

1. 创建脚本

（1）编辑文件 admin-openrc 并添加如下内容：

```
export OS_PROJECT_DOMAIN_NAME=default
export OS_USER_DOMAIN_NAME=default
export OS_PROJECT_NAME=admin
export OS_USERNAME=admin
export OS_PASSWORD=123456
export OS_AUTH_URL=http://controller:35357/v3
export OS_IDENTITY_API_VERSION=3
export OS_IMAGE_API_VERSION=2
```

（2）编辑文件 demo-openrc 并添加如下内容：

```
export OS_PROJECT_DOMAIN_NAME=default
export OS_USER_DOMAIN_NAME=default
export OS_PROJECT_NAME=demo
export OS_USERNAME=demo
export OS_PASSWORD=123456
export OS_AUTH_URL=http://controller:5000/v3
export OS_IDENTITY_API_VERSION=3
export OS_IMAGE_API_VERSION=2
```

2. 使用脚本

使用特定租户和用户运行客户端，可以在运行之前简单地加载相关客户端脚本。例如：

加载 admin-openrc 文件：

```
[root@controller ~]# . admin-openrc
```

请求认证令牌：

```
[root@controller ~]# openstack token issue
+-------------+--------------------------------------------------------------+
| Field       | Value                                                        |
```

```
+-------------+----------------------------------------------------------------+
| expires     | 2016-11-28T17:06:57.565982Z                                    |
| id          | gAAAAABYPFYhFraVYFL0CLQ2uD_gFqp-                                |
|             | LC8btXo6UqfGjBMjpVU2CbR1dgoj8_Ny9w1AQnLbFUBW8nIBq               |
|             | DwkX8T8G4CbkKhPj8WB33ZcNEleaG2ObdkdFjGjCKfuAHEoHPlh6Qx-         |
|             | WhpZ6EmG_tzY8xlcQHNhsetCUSlmIb4kke17y69sEFUCGpA                 |
| project_id  | 524aa64844814bc5bf6f73fd9e416410                                |
| user_id     | c7552cc0ee694bc7b5a855cbbd4165a5                                |
+-------------+----------------------------------------------------------------+
```

2.4 镜像服务

2.4.1 镜像服务概览

镜像服务（glance）允许用户发现、注册和获取虚拟机镜像。它提供了一个 REST API，允许查询虚拟机镜像的元数据并获取一个现存的镜像。可以将虚拟机镜像存储到各种位置，从简单的文件系统到对象存储系统——例如 OpenStack 对象存储，并通过镜像服务使用。

简单来说，使用 file 作为后端配置镜像服务，能够上传并存储在一个托管镜像服务的控制节点目录中。默认情况下，这个目录是 /var/lib/glance/images/。

继续进行之前，确认控制节点的该目录有至少几千兆字节的可用空间。

OpenStack 镜像服务是 IaaS 的核心服务，它接受磁盘镜像或服务器镜像 API 请求和来自终端用户或 OpenStack 计算组件的元数据定义。它也支持包括 OpenStack 对象存储在内的多种类型仓库上的磁盘镜像或服务器镜像存储。

大量周期性进程运行于 OpenStack 镜像服务上以支持缓存。同步复制（Replication）服务保证集群中的一致性和可用性。其他周期性进程包括 auditors、updaters 和 reapers。

OpenStack 镜像服务包括以下组件：

- glance-api：接收镜像 API 的调用，诸如镜像发现、恢复、存储。
- glance-registry：存储、处理和恢复镜像的元数据，元数据包括诸如大小和类型等项。
- glance-registry：私有内部服务，用于服务 OpenStack Image 服务。不要向用户暴露该服务。
- 数据库：存放镜像元数据，用户是可以依据个人喜好选择数据库的，多数的部署使用 MySQL 或 SQLite。

镜像文件的存储仓库：支持多种类型的仓库，它们有普通文件系统、对象存储、RADOS 块设备、HTTP、以及亚马逊 S3。记住，其中一些仓库仅支持只读方式使用。

元数据定义服务：通用的 API，用于为厂商、管理员服务，以及用户自定义元数据。这种元数据可用于不同的资源，例如镜像、工件、卷、配额以及集合。一个定义包括了新属性的键、描述、约束以及可以与之关联的资源类型。

2.4.2 安装和配置

安装和配置镜像服务之前，必须创建一个数据库、服务凭证和 API 端点。

（1）完成下面的步骤以创建数据库：

1）以 root 用户连接到数据库服务器：

```
[root@controller ~]# mysql -u root -p123456
```

2）创建 glance 数据库：

```
MariaDB [(none)]> CREATE DATABASE glance;
Query OK, 1 row affected (0.35 sec)
```

3）对 glance 数据库给予恰当的权限：

```
MariaDB [(none)]> GRANT ALL PRIVILEGES ON glance.* TO 'glance'@'localhost' IDENTIFIED
    BY '123456';
Query OK, 0 rows affected (0.01 sec)

MariaDB [(none)]> GRANT ALL PRIVILEGES ON glance.* TO 'glance'@'%' IDENTIFIED
    BY '123456';
Query OK, 0 rows affected (0.00 sec)
```

（2）获得 admin 凭证来获取只有管理员能执行的命令的访问权限：

```
[root@controller ~]# . admin-openrc
```

（3）创建 glance 用户：

```
[root@controller ~]# openstack user create --domain default --password-prompt glance
User Password:                          // 密码 123456
Repeat User Password: // 密码 123456
+-----------+----------------------------------+
| Field     | Value                            |
+-----------+----------------------------------+
| domain_id | 7174f90333fd46fd8a5f1ca8989df867 |
| enabled   | True                             |
| id        | 10aab1b171a247548be0bef447ca1f27 |
| name      | glance                           |
+-----------+----------------------------------+
```

（4）添加 admin 角色到 glance 用户和 service 项目上：

[root@controller ~]# openstack role add --project service --user glance admin

（5）创建 glance 服务实体：

```
[root@controller ~]# openstack service create --name glance --description "OpenStack Image" image
+-------------+----------------------------------+
| Field       | Value                            |
+-------------+----------------------------------+
| description | OpenStack Image                  |
| enabled     | True                             |
| id          | 00e88017550c470d93caf78dc3de0de3 |
| name        | glance                           |
| type        | image                            |
+-------------+----------------------------------+
```

2.4.3 创建镜像服务的 API 端点

```
[root@controller ~]# openstack endpoint create --region RegionOne image public http://controller:9292
+--------------+----------------------------------+
| Field        | Value                            |
+--------------+----------------------------------+
| enabled      | True                             |
| id           | 2b024219dfa744a7a4c2c8709b668e0d |
| interface    | public                           |
| region       | RegionOne                        |
| region_id    | RegionOne                        |
| service_id   | 00e88017550c470d93caf78dc3de0de3 |
| service_name | glance                           |
| service_type | image                            |
| url          | http://controller:9292           |
+--------------+----------------------------------+
[root@controller ~]# openstack endpoint create --region RegionOne image internal
    http://controller:9292
+--------------+----------------------------------+
| Field        | Value                            |
+--------------+----------------------------------+
| enabled      | True                             |
| id           | 4324b2e2d94f480ea9539e3c1c897d81 |
| interface    | internal                         |
| region       | RegionOne                        |
| region_id    | RegionOne                        |
| service_id   | 00e88017550c470d93caf78dc3de0de3 |
| service_name | glance                           |
```

```
| service_type | image                   |
| url          | http://controller:9292  |
+--------------+-------------------------+
```

[root@controller ~]# openstack endpoint create --region RegionOne image admin http://controller:9292

```
+--------------+----------------------------------+
| Field        | Value                            |
+--------------+----------------------------------+
| enabled      | True                             |
| id           | 04dc5a2607cb42da9fed8e14ac54f3ba |
| interface    | admin                            |
| region       | RegionOne                        |
| region_id    | RegionOne                        |
| service_id   | 00e88017550c470d93caf78dc3de0de3 |
| service_name | glance                           |
| service_type | image                            |
| url          | http://controller:9292           |
+--------------+----------------------------------+
```

2.4.4 安装软件包

[root@controller ~]# yum install openstack-glance

（1）编辑文件 /etc/glance/glance-api.conf 并完成如下操作：

1）在 [database] 部分，配置数据库访问：

[database]
 641 connection = mysql+pymysql://glance:123456@controller/glance

2）在 [keystone_authtoken] 和 [paste_deploy] 部分，配置认证服务访问：

[keystone_authtoken]
1118 auth_uri = http://controller:5000
1119 auth_url = http://controller:35357
1120 memcached_servers = controller:11211
1121 auth_type = password
1122 project_domain_name = default
1123 user_domain_name = default
1124 project_name = service
1125 username = glance
1126 password = 123456

[paste_deploy]
1694 flavor = keystone

3）在 [glance_store] 部分，配置本地文件系统存储和镜像文件位置：

[glance_store]

```
741 stores = file,http
746 default_store = file
1025 filesystem_store_datadir = /var/lib/glance/images
```

（2）编辑文件 /etc/glance/glance-registry.conf 并完成如下操作：

1）在 [database] 部分，配置数据库访问：

```
[database]
 382 connection = mysql+pymysql://glance:123456@controller/glance
```

2）在 [keystone_authtoken] 和 [paste_deploy] 部分，配置认证服务访问：

```
[keystone_authtoken]
 844 auth_uri = http://controller:5000
 845 auth_url = http://controller:35357
 846 memcached_servers = controller:11211
 847 auth_type = password
 848 project_domain_name = default
 849 user_domain_name = default
 850 project_name = service
 851 username = glance
 852 password = 123456

[paste_deploy]
1402 flavor = keystone
```

3）写入镜像服务数据库：

```
[root@controller ~]# su -s /bin/sh -c "glance-manage db_sync" glance
```

忽略输出中任何不推荐使用的信息。

4）启动镜像服务、配置它们随机启动：

```
[root@controller ~]# systemctl enable openstack-glance-api.service openstack-glance-registry.service
[root@controller ~]# systemctl start openstack-glance-api.service openstack-glance-registry.service
```

2.4.5 验证操作

使用 CirrOS 对镜像服务进行验证，CirrOS 是一个小型的 Linux 镜像可以用来帮助进行 OpenStack 部署测试。

关于如何下载和构建镜像的更多信息，参考 OpenStack Virtual Machine Image Guide <http://docs.openstack.org/image-guide/>。关于如何管理镜像的更多信息，参考 OpenStack 用户手册 <http://docs.openstack.org/user-guide/common/cli-manage-images.html>。

（1）获得 admin 凭证来获取只有管理员能执行的命令的访问权限：

```
[root@controller ~]# . admin-openrc
```

（2）下载源镜像：

[root@controller ~]# wget http://download.cirros-cloud.net/0.3.4/cirros-0.3.4-x86_64-disk.img

（3）使用 QCOW2 磁盘格式，bare 容器格式上传镜像到镜像服务并设置公共可见，这样所有的项目都可以访问它：

```
[root@controller ~]# openstack image create "cirros" --file cirros-0.3.4-x86_64-disk.img --disk-format
    qcow2 --container-format bare --public
+------------------+------------------------------------------------------+
| Field            | Value                                                |
+------------------+------------------------------------------------------+
| checksum         | 2a1e86f69acce093d18f21fc0642d431                     |
| container_format | bare                                                 |
| created_at       | 2016-11-28T16:48:54Z                                 |
| disk_format      | qcow2                                                |
| file             | /v2/images/6e1df834-1ee8-482a-91ab-1bba7b846c4d/file |
| id               | 6e1df834-1ee8-482a-91ab-1bba7b846c4d                 |
| min_disk         | 0                                                    |
| min_ram          | 0                                                    |
| name             | cirros                                               |
| owner            | 524aa64844814bc5bf6f73fd9e416410                     |
| protected        | False                                                |
| schema           | /v2/schemas/image                                    |
| size             | 133024                                               |
| status           | active                                               |
| tags             |                                                      |
| updated_at       | 2016-11-28T16:48:56Z                                 |
| virtual_size     | None                                                 |
| visibility       | public                                               |
+------------------+------------------------------------------------------+
```

更多关于命令 glance image-create 的参数信息，请参考 OpenStack Command-Line Interface Reference 中的 Image service command-line client <http://docs.openstack.org/cli-reference/openstack.html#openstack-image-create> 部分。

更多镜像磁盘和容器格式信息，参考 OpenStack 虚拟机镜像指南中的镜像的磁盘及容器格式 <http://docs.openstack.org/image-guide/image-formats.html> 部分。

（4）确认镜像的上传并验证属性：

```
[root@controller ~]# openstack image list
+--------------------------------------+--------+--------+
| ID                                   | Name   | Status |
+--------------------------------------+--------+--------+
| 6e1df834-1ee8-482a-91ab-1bba7b846c4d | cirros | active |
+--------------------------------------+--------+--------+
```

2.5 计算服务

2.5.1 计算服务概览

使用 OpenStack 计算服务来托管和管理云计算系统。OpenStack 计算服务是基础设施即服务（IaaS）系统的主要部分，模块主要由 Python 实现。

OpenStack 计算组件请求 OpenStack Identity 服务进行认证；请求 OpenStack Image 服务提供磁盘镜像；为 OpenStack dashboard 提供用户与管理员接口。磁盘镜像访问限制在项目与用户上；配额以每个项目进行设定（例如，每个项目下可以创建多少实例）。OpenStack组件可以在标准硬件上水平大规模扩展，并且下载磁盘镜像启动虚拟机实例。

OpenStack 计算服务由下列组件构成：

nova-api 服务：接收和响应来自最终用户的计算 API 请求。此服务支持 OpenStack 计算服务 API、Amazon EC2 API 以及特殊的管理 API 用于赋予用户做一些管理的操作。它会强制实施一些规则，发起多数的编排活动，例如运行一个实例。

nova-api-metadata 服务：接受来自虚拟机发送的元数据请求。nova-api-metadata 服务一般在安装 nova-network 服务的多主机模式下使用。更详细的信息，请参考 OpenStack 管理员手册中的链接 Metadata service <http://docs.openstack.org/admin-guide/compute-networking-nova.html#metadata-service>。

nova-compute 服务：一个持续工作的守护进程，通过 Hypervior 的 API 来创建和销毁虚拟机实例。例如：

- XenServer/XCP 的 XenAPI。
- KVM 或 QEMU 的 libvirt。
- VMware 的 VMwareAPI。

过程是比较复杂的。通常情况下，守护进程同意了来自队列的动作请求，转换为一系列的系统命令如启动一个 KVM 实例，然后，到数据库中更新它的状态。

- nova-scheduler 服务：拿到一个来自队列请求虚拟机实例，然后决定哪台计算服务器主机来运行它。

- nova-conductor 模块：媒介作用于 nova-compute 服务与数据库之间。它排除了由 nova-compute 服务对云数据库的直接访问。nova-conductor 模块可以水平扩展。但是，不要将它部署在运行 nova-conductor 服务的主机节点上。参考 Configuration Reference Guide <http://docs.openstack.org/mitaka/config-reference/compute/conductor.html>。

- nova-cert 模块：服务器守护进程向 Nova Cert 服务提供 X509 证书，用来为 euca-bundle-image 生成证书。仅仅是在 EC2 API 的请求中使用。

- nova-network worker 守护进程：与 nova-compute 服务类似，从队列中接受网络任务，并且操作网络。执行任务例如创建桥接的接口或者改变 IPtables 的规则。

- nova-consoleauth 守护进程：授权控制台代理所提供的用户令牌。详情可查看 nova-novncproxy 和 nova-xvpvncproxy。该服务必须为控制台代理运行才可奏效。在集群配置中可以运行二者中任一代理服务而非仅运行一个 nova-consoleauth 服务。更多关于 nova-consoleauth 的信息，请查看 About nova-consoleauth <http://docs.openstack.org/admin-guide/compute-remote-console-access.html#about-nova-consoleauth>。
- nova-novncproxy 守护进程：提供一个代理，用于访问正在运行的实例，通过 VNC 协议，支持基于浏览器的 novnc 客户端。
- nova-spicehtml5proxy 守护进程：提供一个代理，用于访问正在运行的实例，通过 SPICE 协议，支持基于浏览器的 HTML5 客户端。
- nova-xvpvncproxy 守护进程：提供一个代理，用于访问正在运行的实例，通过 VNC 协议，支持 OpenStack 特定的 Java 客户端。
- nova-cert 守护进程：X509 证书。
- nova 客户端：用于用户作为租户管理员或最终用户来提交命令。
- 队列：一个在守护进程间传递消息的中央集线器。常见实现有 RabbitMQ <http://www.rabbitmq.com/>，以及如 Zero MQ <http://www.zeromq.org/> 等 AMQP 消息队列。
- SQL 数据库：存储构建时和运行时的状态，为云基础设施，包括有：可用实例类型、使用中的实例、可用网络、项目。

理论上，OpenStack 计算可以支持任何 SQL-Alchemy 所支持的后端数据库，通常使用 SQLite3 来做测试开发工作，MySQL 和 PostgreSQL 作生产环境。

2.5.2 安装并配置控制节点

这个部分将描述如何在控制节点上安装和配置 Compute 服务，即 nova。

在安装和配置 Compute 服务前，必须创建数据库服务的凭据以及 API 端点。

（1）以 root 用户连接到数据库服务器：

```
[root@controller ~]# mysql -u root -p123456
```

1）创建 nova_api 和 nova 数据库：

```
MariaDB [(none)]> CREATE DATABASE nova_api;
Query OK, 1 row affected (0.01 sec)
MariaDB [(none)]> CREATE DATABASE nova;
Query OK, 1 row affected (0.00 sec)
```

2）对数据库进行正确的授权：

```
MariaDB [(none)]> GRANT ALL PRIVILEGES ON nova_api.* TO 'nova'@'localhost' IDENTIFIED
    BY '123456';
Query OK, 0 rows affected (0.01 sec)
```

```
MariaDB [(none)]> GRANT ALL PRIVILEGES ON nova_api.* TO 'nova'@'%' IDENTIFIED
    BY '123456';
Query OK, 0 rows affected (0.00 sec)

MariaDB [(none)]> GRANT ALL PRIVILEGES ON nova.* TO 'nova'@'localhost' IDENTIFIED
    BY '123456';
Query OK, 0 rows affected (0.00 sec)

MariaDB [(none)]> GRANT ALL PRIVILEGES ON nova.* TO 'nova'@'%' IDENTIFIED
    BY '123456';
Query OK, 0 rows affected (0.00 sec)
```

（2）获得 admin 凭证来获取只有管理员能执行的命令的访问权限：

```
[root@controller ~]# . admin-openrc
```

（3）创建 nova 用户：

```
[root@controller ~]# openstack user create --domain default --password-prompt nova
User Password:                        // 密码 123456
Repeat User Password:                 // 密码 123456
+-----------+----------------------------------+
| Field     | Value                            |
+-----------+----------------------------------+
| domain_id | 7174f90333fd46fd8a5f1ca8989df867 |
| enabled   | True                             |
| id        | 3570fb156f6f454aa0f49bcaa0d80baf |
| name      | nova                             |
+-----------+----------------------------------+
```

（4）给 nova 用户添加 admin 角色：

```
[root@controller ~]# openstack role add --project service --user nova admin
```

（5）创建 nova 服务实体：

```
[root@controller ~]# openstack service create --name nova --description "OpenStack Compute"
    compute
+-------------+----------------------------------+
| Field       | Value                            |
+-------------+----------------------------------+
| description | OpenStack Compute                |
| enabled     | True                             |
| id          | a69249de69cf47aab652173f3982e2b7 |
| name        | nova                             |
| type        | compute                          |
+-------------+----------------------------------+
```

（6）创建 Compute 服务 API 端点：

```
[root@controller ~]# openstack endpoint create --region RegionOne compute public
    http://controller:8774/v2.1/%\(tenant_id\)s
 compute admin http://controller:8774/v2.1/%\(tenant_id\)s
+--------------+------------------------------------------+
| Field        | Value                                    |
+--------------+------------------------------------------+
| enabled      | True                                     |
| id           | a1bd86e706e44bcd889f53e487e07a34         |
| interface    | public                                   |
| region       | RegionOne                                |
| region_id    | RegionOne                                |
| service_id   | a69249de69cf47aab652173f3982e2b7         |
| service_name | nova                                     |
| service_type | compute                                  |
| url          | http://controller:8774/v2.1/%(tenant_id)s |
+--------------+------------------------------------------+
[root@controller ~]# openstack endpoint create --region RegionOne compute internal
    http://controller:8774/v2.1/%\(tenant_id\)s
+--------------+------------------------------------------+
| Field        | Value                                    |
+--------------+------------------------------------------+
| enabled      | True                                     |
| id           | c856873f0bfa4e27961d696eced2011f         |
| interface    | internal                                 |
| region       | RegionOne                                |
| region_id    | RegionOne                                |
| service_id   | a69249de69cf47aab652173f3982e2b7         |
| service_name | nova                                     |
| service_type | compute                                  |
| url          | http://controller:8774/v2.1/%(tenant_id)s |
+--------------+------------------------------------------+
[root@controller ~]# openstack endpoint create --region RegionOne compute admin
    http://controller:8774/v2.1/%\(tenant_id\)s
+--------------+------------------------------------------+
| Field        | Value                                    |
+--------------+------------------------------------------+
| enabled      | True                                     |
| id           | ca7f0a957a554926a23544d295bffda3         |
| interface    | admin                                    |
| region       | RegionOne                                |
| region_id    | RegionOne                                |
| service_id   | a69249de69cf47aab652173f3982e2b7         |
| service_name | nova                                     |
```

```
| service_type | compute                                    |
| url          | http://controller:8774/v2.1/%(tenant_id)s |
+--------------+--------------------------------------------+
```

（7）安装软件包：

```
[root@controller ~]# yum -y install openstack-nova-api openstack-nova-conductor openstack-nova
   -console openstack-nova-novncproxy
 openstack-nova-scheduler
```

编辑 /etc/nova/nova.conf 文件并完成下面的操作：

1）在 [DEFAULT] 部分，只启用计算和元数据 API：

```
[DEFAULT]
 267 enabled_apis=osapi_compute,metadata
```

2）在 [api_database] 和 [database] 部分，配置数据库的连接：

```
[api_database]
2168 connection = mysql+pymysql://nova:123456@controller/nova_api

[database]
3224 connection = mysql+pymysql://nova:123456@controller/nova
```

3）在 [DEFAULT] 和 [oslo_messaging_rabbit] 部分，配置 RabbitMQ 消息队列访问：

```
[DEFAULT]
2119 rpc_backend=rabbit

[oslo_messaging_rabbit]
4448 rabbit_host=controller
4466 rabbit_userid=openstack
4470 rabbit_password=123456
```

4）在 [DEFAULT] 和 [keystone_authtoken] 部分，配置认证服务访问：

```
[DEFAULT]
 382 auth_strategy=keystone

[keystone_authtoken]
3528 auth_uri = http://controller:5000
3529 auth_url = http://controller:35357
3530 memcached_servers = controller:11211
3531 auth_type = password
3532 project_domain_name = default
3533 user_domain_name = default
3534 project_name = service
3535 username = nova
3536 password = 123456
```

5）在 [DEFAULT] 部分，配置 my_ip 来使用控制节点的管理接口的 IP 地址：

```
[DEFAULT]
143 my_ip=10.10.10.10
```

6）在 [DEFAULT] 部分，使能 Networking 服务：

```
[DEFAULT]
1684 use_neutron=True
1561 firewall_driver = nova.virt.firewall.NoopFirewallDriver
```

默认情况下，计算服务使用内置的防火墙服务。由于网络服务包含了防火墙服务，必须使用 nova.virt.firewall.NoopFirewallDriver 防火墙服务来禁用掉计算服务内置的防火墙服务。

7）在 [vnc] 部分，配置 VNC 代理使用控制节点的管理接口 IP 地址：

```
[vnc]
5392 enabled=True
5424 vncserver_listen=$my_ip                    // 修改成 0.0.0.0
5448 vncserver_proxyclient_address=$my_ip       // 修改成具体 IP 地址
```

8）在 [glance] 部分，配置镜像服务 API 的位置：

```
[glance]
3353 api_servers=http://controller:9292
```

9）在 [oslo_concurrency] 部分，配置锁路径：

```
[oslo_concurrency]
4304 lock_path=/var/lib/nova/tmp
```

10）同步 Compute 数据库：

```
[root@controller ~]# su -s /bin/sh -c "nova-manage api_db sync" nova
[root@controller ~]# su -s /bin/sh -c "nova-manage db sync" nova
```

（8）完成安装。

启动 Compute 服务并将其设置为随系统启动：

```
[root@controller ~]# systemctl enable openstack-nova-api.service  openstack-nova-consoleauth.
    service openstack-nova-scheduler.s
ervice  openstack-nova-conductor.service openstack-nova-novncproxy.service
[root@controller ~]# systemctl start openstack-nova-api.service   openstack-nova-consoleauth.service
    openstack-nova-scheduler.service   openstack-nova-conductor.service openstack-nova-novncproxy.
    service
```

2.5.3 安装和配置计算节点

本节描述如何在计算节点上安装并配置计算服务。

1. 安装软件包

[root@compute ~]# yum -y install centos-release-openstack-mitaka

[root@compute ~]# yum -y install openstack-nova-compute

编辑 /etc/nova/nova.conf 文件并完成下面的操作：

（1）在 [DEFAULT] 和 [oslo_messaging_rabbit] 部分，配置 RabbitMQ 消息队列的连接：

[DEFAULT]
2119 rpc_backend=rabbit

[oslo_messaging_rabbit]
4449 rabbit_host=controller
4467 rabbit_userid=openstack
4471 rabbit_password=123456

（2）在 [DEFAULT] 和 [keystone_authtoken] 部分，配置认证服务访问：

[DEFAULT]
 382 auth_strategy=keystone

[keystone_authtoken]
3529 auth_uri = http://controller:5000
3530 auth_url = http://controller:35357
3531 memcached_servers = controller:11211
3532 auth_type = password
3533 project_domain_name = default
3534 user_domain_name = default
3535 project_name = service
3536 username = nova
3537 password = 123456

（3）在 [DEFAULT] 部分，配置 my_ip 选项：

[DEFAULT]
143 my_ip=10.10.10.12   // 计算节点上管理网络接口的 IP 地址

（4）在 [DEFAULT] 部分，使能 Networking 服务：

[DEFAULT]
1684 use_neutron=True
1561 firewall_driver = nova.virt.firewall.NoopFirewallDriver

Compute 使用内置的防火墙服务。由于 Networking 包含了防火墙服务，所以必须通过使用 nova.virt.firewall.NoopFirewallDriver 来去除 Compute 内置的防火墙服务。

(5) 在 [vnc] 部分，启用并配置远程控制台访问：

```
[vnc]
5383 enabled=true
5425 vncserver_listen=0.0.0.0
5449 vncserver_proxyclient_address=$my_ip         // 修改成具体 IP 地址
5530 novncproxy_base_url = http://controller:6080/vnc_auto.html
```

如果运行浏览器的主机无法解析 controller 主机名，可以将 controller 替换为控制节点管理网络的 IP 地址。

(6) 在 [glance] 部分，配置镜像服务 API 的位置：

```
[glance]
3354 api_servers=http://controller:9292
```

(7) 在 [oslo_concurrency] 部分，配置锁路径：

```
[oslo_concurrency]
4305 lock_path=/var/lib/nova/tmp
```

2. 完成安装

(1) 确定计算节点是否支持虚拟机的硬件加速。

```
[root@compute ~]# egrep -c '(vmx|svm)' /proc/cpuinfo
1
```

如果这个命令返回了 one or greater 的值，那么计算节点支持硬件加速且不需要额外的配置。

如果这个命令返回了 zero 值，那么计算节点不支持硬件加速，必须配置 libvirt 来使用 QEMU 去代替 KVM。

在 /etc/nova/nova.conf 文件的 [libvirt] 部分做出如下的编辑：

```
[libvirt]
3681 virt_type=qemu
```

(2) 启动计算服务及其依赖，并将其配置为随系统自动启动：

```
[root@compute ~]# systemctl enable libvirtd.service openstack-nova-compute.service
[root@compute ~]# systemctl start libvirtd.service openstack-nova-compute.service
```

2.5.4 验证操作

在控制节点上执行这些命令：

(1) 获得 admin 凭证来获取只有管理员能执行的命令的访问权限：

```
[root@controller ~]# . admin-openrc
```

(2) 列出服务组件，以验证是否成功启动并注册了每个进程：

```
[root@controller ~]# openstack compute service list
+----+------------------+------------+----------+---------+-------+----------------------------+
| Id | Binary           | Host       | Zone     | Status  | State | Updated At                 |
+----+------------------+------------+----------+---------+-------+----------------------------+
|  1 | nova-consoleauth | controller | internal | enabled | up    | 2016-11-28T17:47:55.000000 |
|  2 | nova-conductor   | controller | internal | enabled | up    | 2016-11-28T17:47:55.000000 |
|  3 | nova-scheduler   | controller | internal | enabled | up    | 2016-11-28T17:47:55.000000 |
|  6 | nova-compute     | compute    | nova     | enabled | up    | 2016-11-28T17:47:52.000000 |
+----+------------------+------------+----------+---------+-------+----------------------------+
```

该输出应该显示三个服务组件在控制节点上启用，一个服务组件在计算节点上启用。

2.6 Networking 服务

2.6.1 网络服务概览

OpenStack Networking（neutron），允许创建、插入接口设备，这些设备由其他的 OpenStack 服务管理。插件式的实现可以容纳不同的网络设备和软件，为 OpenStack 架构与部署提供了灵活性。

它包含下列组件：

- neutron-server：接收和路由 API 请求到合适的 OpenStack 网络插件，以达到预想的目的。
- OpenStack 网络插件和代理：插拔端口，创建网络和子网，以及提供 IP 地址，这些插件和代理依赖于供应商和技术而不同，OpenStack 网络基于插件和代理为 Cisco 虚拟和物理交换机、NEC OpenFlow 产品、Open vSwitch、Linux bridging 以及 VMware NSX 产品穿线搭桥。

常见的代理 L3（3 层）、DHCP（动态主机 IP 地址），以及插件代理。

- 消息队列：大多数的 OpenStack Networking 安装都会用到，用于在 neutron-server 和各种各样的代理进程间路由信息。也为某些特定的插件扮演数据库的角色，以存储网络状态。OpenStack 网络主要和 OpenStack 计算交互，以提供网络连接到它的实例。
- 网络（neutron）概念：OpenStack 网络（neutron）管理 OpenStack 环境中所有虚拟网络基础设施（VNI）、物理网络基础设施（PNI）的接入层。OpenStack 网络允许租户创建包括像 firewall、:term:`load balancer` 和 :term:`virtual private network (VPN)` 等这样的高级虚拟网络拓扑。

网络服务提供网络、子网以及路由这些对象的抽象概念。每个抽象概念都有自己的功能，可以模拟对应的物理设备：网络包括子网，路由在不同的子网和网络间进行路由转发。

对于任意一个给定的网络都必须包含至少一个外部网络。不像其他的网络那样，外部网络不仅仅是一个定义的虚拟网络。相反，它代表了一种 OpenStack 安装之外的能从物理的、外部的网络访问的视图。外部网络上的 IP 地址可供外部网络上的任意的物理设备访问。

外部网络之外，任何 Networking 设置拥有一个或多个内部网络。这些软件定义的网络直接连接到虚拟机。仅仅在给定网络上的虚拟机，或那些通过接口连接到相近路由的子网上的虚拟机，能直接访问连接到那个网络上的虚拟机。

如果外部网络想要访问实例或者实例想要访问外部网络，那么网络之间的路由就是必要的了。每一个路由都配有一个网关用于连接到外部网络，以及一个或多个连接到内部网络的接口。就像一个物理路由一样，子网可以访问同一个路由上其他子网中的机器，并且机器也可以通过访问路由的网关访问外部网络。

另外，可以将外部网络的 IP 地址分配给内部网络的端口。不管什么时候一旦有连接连接到子网，那个连接就被称作端口。可以给实例的端口分配外部网络的 IP 地址。通过这种方式，外部网络上的实体可以访问实例。

网络服务同样支持安全组。安全组允许管理员在安全组中定义防火墙规则。一个实例可以属于一个或多个安全组，网络为这个实例配置这些安全组中的规则，阻止或者开启端口，端口范围或者通信类型。

每一个 Networking 使用的插件都有其自有的概念。虽然对操作 VNI 和 OpenStack 环境不是至关重要的，但理解这些概念能帮助设置 Networking。所有的 Networking 安装使用了一个核心插件和一个安全组插件（或仅是空操作安全组插件）。另外，防火墙即服务（FWaaS）和负载均衡即服务（LBaaS）插件是可用的。

2.6.2 安装并配置控制节点

（1）在配置 OpenStack 网络（neutron）服务之前，必须为其创建一个数据库、服务凭证和 API 端点。

1) 以 root 用户连接到数据库服务器：

```
[root@controller ~]# mysql -u root -p123456
```

- 创建 neutron 数据库：

```
MariaDB [(none)]> CREATE DATABASE neutron;
Query OK, 1 row affected (0.00 sec)
```

- 对 neutron 数据库给予合适的访问权限，使用合适的密码替换 NEUTRON_DBPASS：

```
MariaDB [(none)]> GRANT ALL PRIVILEGES ON neutron.* TO 'neutron'@'localhost' IDENTIFIED BY '123456';
Query OK, 0 rows affected (0.01 sec)
MariaDB [(none)]> GRANT ALL PRIVILEGES ON neutron.* TO 'neutron'@'%' IDENTIFIED
```

```
    BY '123456';
Query OK, 0 rows affected (0.00 sec)
```

2）获得 admin 凭证来获取只有管理员能执行的命令的访问权限：

```
[root@controller ~]# . admin-openrc
```

3）要创建服务证书，完成这些步骤：

- 创建 neutron 用户：

```
[root@controller ~]# openstack user create --domain default --password-prompt neutron
User Password:                    // 密码 123456
Repeat User Password:             // 密码 123456
+-----------+----------------------------------+
| Field     | Value                            |
+-----------+----------------------------------+
| domain_id | 7174f90333fd46fd8a5f1ca8989df867 |
| enabled   | True                             |
| id        | e55b9580b03e4cb4a3c0640fe43a55c5 |
| name      | neutron                          |
+-----------+----------------------------------+
```

- 添加 admin 角色到 neutron 用户：

```
[root@controller ~]# openstack role add --project service --user neutron admin
```

- 创建 neutron 服务实体：

```
[root@controller ~]# openstack service create --name neutron --description "OpenStack Networking"
    network
+-------------+----------------------------------+
| Field       | Value                            |
+-------------+----------------------------------+
| description | OpenStack Networking             |
| enabled     | True                             |
| id          | abc28f0359924f67a842676c149175a1 |
| name        | neutron                          |
| type        | network                          |
+-------------+----------------------------------+
```

4）创建网络服务 API 端点：

```
[root@controller ~]# openstack endpoint create --region RegionOne network public
    http://controller:9696
+--------------+----------------------------------+
| Field        | Value                            |
+--------------+----------------------------------+
| enabled      | True                             |
| id           | 449c050cf15e4aefbc8d71ae26cfdda1 |
| interface    | public                           |
| region       | RegionOne                        |
```

```
| region_id    | RegionOne                        |
| service_id   | abc28f0359924f67a842676c149175a1 |
| service_name | neutron                          |
| service_type | network                          |
| url          | http://controller:9696           |
+--------------+----------------------------------+
```

[root@controller ~]# openstack endpoint create --region RegionOne network internal http://controller:9696

```
+--------------+----------------------------------+
| Field        | Value                            |
+--------------+----------------------------------+
| enabled      | True                             |
| id           | 0e81f9b0d8f84e4e8bd6e5ee6068cd3b |
| interface    | internal                         |
| region       | RegionOne                        |
| region_id    | RegionOne                        |
| service_id   | abc28f0359924f67a842676c149175a1 |
| service_name | neutron                          |
| service_type | network                          |
| url          | http://controller:9696           |
+--------------+----------------------------------+
```

[root@controller ~]# openstack endpoint create --region RegionOne network admin http://controller:9696

```
+--------------+----------------------------------+
| Field        | Value                            |
+--------------+----------------------------------+
| enabled      | True                             |
| id           | 4636de62d51147789a8c4cb4b4d0f05f |
| interface    | admin                            |
| region       | RegionOne                        |
| region_id    | RegionOne                        |
| service_id   | abc28f0359924f67a842676c149175a1 |
| service_name | neutron                          |
| service_type | network                          |
| url          | http://controller:9696           |
+--------------+----------------------------------+
```

（2）配置网络选项：

可以部署网络服务使用选项1和选项2两种架构中的一种来部署网络服务。

选项1采用尽可能简单的架构进行部署，只支持实例连接到公有网络（外部网络）。没有私有网络（个人网络），路由器以及浮动IP地址。只有admin或者其他特权用户才可以管理公有网络。

选项2在选项1的基础上多了layer-3服务，支持实例连接到私有网络。demo或者其他没有特权的用户可以管理自己的私有网络，包含连接公网和私网的路由器。另外，浮动IP地址可以让实例使用私有网络连接到外部网络，例如互联网。

典型的私有网络一般使用覆盖网络。覆盖网络，例如 VXLAN 包含了额外的数据头，这些数据头增加了开销，减少了有效内容和用户数据的可用空间。在不了解虚拟网络架构的情况下，实例尝试用以太网最大传输单元（MTU）1500 字节发送数据包。网络服务会自动给实例提供正确的 MTU 的值通过 DHCP 的方式。但是，一些云镜像并没有使用 DHCP 或者忽视了 DHCP MTU 选项，要求使用元数据或者脚本来进行配置。

网络选项 1：公共网络

在 controller 节点上安装并配置网络组件

1）安装组件。

```
[root@controller ~]# yum install openstack-neutron openstack-neutron-ml2 openstack-neutron-linuxbridge ebtables -y
```

2）配置服务组件。

Networking 服务器组件的配置包括数据库、认证机制、消息队列、拓扑变化通知和插件。

编辑 /etc/neutron/neutron.conf 文件并完成如下操作：

- 在 [database] 部分，配置数据库访问：

```
[root@controller ~]# vim /etc/neutron/neutron.conf
[database]
 684 connection = mysql+pymysql://neutron:123456@controller/neutron
```

- 在 [DEFAULT] 部分，启用 ML2 插件并禁用其他插件：

```
[DEFAULT]
 30 core_plugin = ml2
 33 service_plugins =                  // 删除 router 关键词（vlan 网络 +vxlan 网络）
```

- 在 [DEFAULT] 部分，配置 RabbitMQ 消息队列的连接：

```
[DEFAULT]
 511 rpc_backend = rabbit
1194 rabbit_host = controller
1212 rabbit_userid = openstack
1216 rabbit_password = 123456
```

- 在 [DEFAULT] 和 [keystone_authtoken] 部分，配置认证服务访问：

```
[DEFAULT]
  27 auth_strategy = keystone

[keystone_authtoken]
 768 auth_uri = http://controller:5000
 769 auth_url = http://controller:35357
 770 memcached_servers = controller:11211
 771 auth_type = password
 772 project_domain_name = default
```

```
773 user_domain_name = default
774 project_name = service
775 username = neutron
776 password = 123456
```

- 在 [DEFAULT] 和 [nova] 部分，配置网络服务来通知计算节点的网络拓扑变化：

```
[DEFAULT]
137 notify_nova_on_port_status_changes = true
141 notify_nova_on_port_data_changes = true

[nova]
958 auth_url = http://controller:35357
959 auth_type = password
960 project_domain_name = default
961 user_domain_name = default
962 region_name = RegionOne
963 project_name = service
964 username = nova
965 password = 123456
```

- 在 [oslo_concurrency] 部分，配置锁路径：

```
[oslo_concurrency]
1059 lock_path = /var/lib/neutron/tmp
```

3）配置 Modular Layer 2（ML2）插件。
ML2 插件使用 Linuxbridge 机制来为实例创建 layer-2 虚拟网络基础设施。
编辑 /etc/neutron/plugins/ml2/ml2_conf.ini 文件并完成以下操作：

- 在 [ml2] 部分，启用 flat 和 VLAN 网络：

```
[root@controller ~]# vim /etc/neutron/plugins/ml2/ml2_conf.ini
[ml2]
107 type_drivers = flat,vlan
```

- 在 [ml2] 部分，禁用私有网络：

```
112 tenant_network_types =                    // 删除 flat,vlan
// 这个是租户网络，采用网络 1 的话，是 vlan 网络，所有的租户都在同一个网络里，比如用户 a
   和用户 b 之间的内网就可以互通；这里采用网络 1，所以这个是不用填的
```

- 在 [ml2] 部分，启用 Linuxbridge 机制：

```
116 mechanism_drivers = linuxbridge
```

- 在 [ml2] 部分，启用端口安全扩展驱动：

```
121 extension_drivers = port_security
```

- 在 [ml2_type_flat] 部分，配置公共虚拟网络为 flat 网络：

```
[ml2_type_flat]
```

153 flat_networks = provider

- 在 [securitygroup] 部分，启用 ipset 增加安全组规则的高效性：

[securitygroup]
230 enable_ipset = true

4）配置 Linuxbridge 代理。

Linuxbridge 代理为实例建立 layer-2 虚拟网络并且处理安全组规则。

编辑 /etc/neutron/plugins/ml2/linuxbridge_agent.ini 文件并且完成以下操作：

- 在 [linux_bridge] 部分，将公共虚拟网络和公共物理网络接口对应起来：

[root@controller ~]# vim /etc/neutron/plugins/ml2/linuxbridge_agent.ini
[linux_bridge]
138 physical_interface_mappings = provider:eno33554952

将 PUBLIC_INTERFACE_NAME 替换为底层的物理公共网络接口。

- 在 [vxlan] 部分，禁止 VXLAN 覆盖网络：

[vxlan]
171 enable_vxlan = False

- 在 [securitygroup] 部分，启用安全组并配置 Linuxbridge iptables firewall driver：

[securitygroup]
156 enable_security_group = true
151 firewall_driver = neutron.agent.linux.iptables_firewall.IptablesFirewallDrive

5）配置 DHCP 代理。

The DHCP agent provides DHCP services for virtual networks.

编辑 /etc/neutron/dhcp_agent.ini 文件并完成下面的操作：

在 [DEFAULT] 部分，配置 Linuxbridge 驱动接口，DHCP 驱动并启用隔离元数据，这样在公共网络上的实例就可以通过网络来访问元数据：

[DEFAULT]
 23 interface_driver = neutron.agent.linux.interface.BridgeInterfaceDriver
 39 dhcp_driver = neutron.agent.linux.dhcp.Dnsmasq
 48 enable_isolated_metadata = True

6）配置元数据代理。

metadata agent 负责提供配置信息，例如：访问实例的凭证。

编辑 /etc/neutron/metadata_agent.ini 文件并完成以下操作：

- 在 [DEFAULT] 部分，配置元数据主机以及共享密码：

[root@controller ~]# vim /etc/neutron/metadata_agent.ini
[DEFAULT]
nova_metadata_ip = controller
metadata_proxy_shared_secret = 123456

用为元数据代理设置的密码替换 METADATA_SECRET。

7）为计算节点配置网络服务。

编辑 /etc/nova/nova.conf 文件并完成以下操作：

- 在 [neutron] 部分，配置访问参数，启用元数据代理并设置密码：

```
[neutron]
4153 url = http://controller:9696
4154 auth_url = http://controller:35357
4155 auth_type = password
4156 project_domain_name = default
4157 user_domain_name = default
4158 region_name = RegionOne
4159 project_name = service
4160 username = neutron
4161 password = 123456

4163 service_metadata_proxy = True
4164 metadata_proxy_shared_secret = 123456
```

8）完成安装。

- 网络服务初始化脚本需要一个超链接 /etc/neutron/plugin.ini 指向 ML2 插件配置文件 /etc/neutron/plugins/ml2/ml2_conf.ini。如果超链接不存在，使用下面的命令创建它：

```
[root@controller ~]# ln -s /etc/neutron/plugins/ml2/ml2_conf.ini /etc/neutron/plugin.ini
```

- 同步数据库：

```
[root@controller ~]# su -s /bin/sh -c "neutron-db-manage --config-file /etc/neutron/neutron.conf --config-file /etc/neutron/plugins/ml2/ml2_conf.ini upgrade head" neutron
```

- 重启计算 API 服务：

```
[root@controller ~]# systemctl restart openstack-nova-api.service
```

- 当系统启动时，启动 Networking 服务并配置它启动。

对于两种网络选项：

```
[root@controller ~]# systemctl enable neutron-server.service neutron-linuxbridge-agent.service neutron-dhcp-agent.service neutron-metadata-agent.service

[root@controller ~]# systemctl start neutron-server.service neutron-linuxbridge-agent.service neutron-dhcp-agent.service neutron-metadata-agent.service
```

2.6.3 安装和配置计算节点

计算节点处理实例的连接和安全组。

1. 安装组件

```
[root@compute ~]# yum install openstack-neutron-linuxbridge ebtables ipset
```

2. 配置通用组件

Networking 通用组件的配置包括认证机制、消息队列和插件。

编辑 /etc/neutron/neutron.conf 文件并完成如下操作：

在 [database] 部分，注释所有 connection 项，因为计算节点不直接访问数据库。

在 [DEFAULT] 和 [oslo_messaging_rabbit] 部分，配置 "RabbitMQ" 消息队列的连接：

```
[DEFAULT]
 511 rpc_backend = rabbit

[oslo_messaging_rabbit]
1194 rabbit_host = controller
1212 rabbit_userid = openstack
1216 rabbit_password = 123456
```

在 [DEFAULT] 和 [keystone_authtoken] 部分，配置认证服务访问：

```
[DEFAULT]
 27 auth_strategy = keystone

[keystone_authtoken]
 768 auth_uri = http://controller:5000
 769 auth_url = http://controller:35357
 770 memcached_servers = controller:11211
 771 auth_type = password
 772 project_domain_name = default
 773 user_domain_name = default
 774 project_name = service
 775 username = neutron
 776 password = 123456
```

在 [oslo_concurrency] 部分，配置锁路径：

```
[oslo_concurrency]
1051 lock_path = /var/lib/neutron/tmp
```

3. 配置网络选项

选择与之前在控制节点上选择的相同的网络选项。之后，回到这里并进行下一步：为计算节点配置网络服务。

网络选项 1：公共网络

在计算节点上配置网络组件

（1）配置 Linuxbridge 代理

Linuxbridge 代理为实例建立 layer-2 虚拟网络并且处理安全组规则。

编辑 /etc/neutron/plugins/ml2/linuxbridge_agent.ini 文件并且完成以下操作：

1）在 [linux_bridge] 部分，将公共虚拟网络和公共物理网络接口对应起来：

[linux_bridge]
138 physical_interface_mappings = provider:eno33554952

2）在 [vxlan] 部分，禁止 VXLAN 覆盖网络：

[vxlan]
171 enable_vxlan = False

3）在 [securitygroup] 部分，启用安全组并配置 Linuxbridge iptables firewall driver：

[securitygroup]
156 enable_security_group = true
151 firewall_driver = neutron.agent.linux.iptables_firewall.IptablesFirewallDriver

（2）为计算节点配置网络服务

编辑 /etc/nova/nova.conf 文件并完成下面的操作：

在 [neutron] 部分，配置访问参数：

[neutron]
4154 url = http://controller:9696
4155 auth_url = http://controller:35357
4156 auth_type = password
4157 project_domain_name = default
4158 user_domain_name = default
4159 region_name = RegionOne
4160 project_name = service
4161 username = neutron
4162 password = 123456

（3）完成安装

1）重启计算服务：

[root@compute ~]# systemctl restart openstack-nova-compute.service

2）启动 Linuxbridge 代理并配置它开机自启动：

[root@compute ~]# systemctl enable neutron-linuxbridge-agent.service
[root@compute ~]# systemctl start neutron-linuxbridge-agent.service

2.6.4 验证操作

在控制节点上执行这些命令。

（1）获得 admin 凭证来获取只有管理员能执行的命令的访问权限：

[root@controller ~]# . admin-openrc

（2）列出加载的扩展来验证 neutron-server 进程是否正常启动：

```
[root@controller ~]# neutron ext-list
+---------------------------+---------------------------------------------+
| alias                     | name                                        |
+---------------------------+---------------------------------------------+
| default-subnetpools       | Default Subnetpools                         |
| network-ip-availability   | Network IP Availability                     |
| network_availability_zone | Network Availability Zone                   |
| auto-allocated-topology   | Auto Allocated Topology Services            |
| ext-gw-mode               | Neutron L3 Configurable external gateway mode |
| binding                   | Port Binding                                |
| agent                     | agent                                       |
| subnet_allocation         | Subnet Allocation                           |
| l3_agent_scheduler        | L3 Agent Scheduler                          |
| tag                       | Tag support                                 |
| external-net              | Neutron external network                    |
| net-mtu                   | Network MTU                                 |
| availability_zone         | Availability Zone                           |
| quotas                    | Quota management support                    |
| l3-ha                     | HA Router extension                         |
| provider                  | Provider Network                            |
| multi-provider            | Multi Provider Network                      |
| address-scope             | Address scope                               |
| extraroute                | Neutron Extra Route                         |
| timestamp_core            | Time Stamp Fields addition for core resources |
| router                    | Neutron L3 Router                           |
| extra_dhcp_opt            | Neutron Extra DHCP opts                     |
| security-group            | security-group                              |
| dhcp_agent_scheduler      | DHCP Agent Scheduler                        |
| router_availability_zone  | Router Availability Zone                    |
| rbac-policies             | RBAC Policies                               |
| standard-attr-description | standard-attr-description                   |
| port-security             | Port Security                               |
| allowed-address-pairs     | Allowed Address Pairs                       |
| dvr                       | Distributed Virtual Router                  |
+---------------------------+---------------------------------------------+
```

网络选项 1：公共网络

列出代理以验证启动 neutron 代理是否成功：

```
[root@controller ~]# neutron agent-list
+----------------------+------------+------------+-------------------+-------+------------------+-------------------+
| id                   | agent_type | host       | availability_zone | alive | admin_state_up   | binary            |
+----------------------+------------+------------+-------------------+-------+------------------+-------------------+
| 08831df1-aac8-4be7   | DHCP agent | controller | nova              | :-)   | True             | neutron-dhcp-agent |
```

	-96fa-86b06fa6d785						
	4814f607-0236-44f0	Linux bridge agent	compute		:-)	True	neutron-linuxbridge-
	-b91b-6624ec63a950						agent
	da794530-c5e8-4d5c-	Metadata agent	controller		:-)	True	neutron-metadata-
	bd47-63a03fa10f6c						agent
+---------------------+--------------------+------------+---------------+-------+-------------+----------------------+

输出结果应该包括控制节点上的三个代理和每个计算节点上的一个代理。

现在 OpenStack 环境包含了启动一个基本实例所必须的核心组件。可以参考 launch-instance 或者添加更多的 OpenStack 服务到环境中。

2.7 Dashboard

Dashboard（horizon）是一个 Web 接口，使得云平台管理员以及用户可以管理不同的 OpenStack 资源以及服务。

这个部署示例使用的是 Apache Web 服务器。

这个部分将描述如何在控制节点上安装和配置仪表板。

1. 安装并配置组件

（1）安装软件包：

```
[root@controller ~]# yum install openstack-dashboard -y
```

（2）编辑文件 /etc/openstack-dashboard/local_settings 并完成如下操作：

在 controller 节点上配置仪表盘以使用 OpenStack 服务：

```
[root@controller ~]# vim /etc/openstack-dashboard/local_settings
158 OPENSTACK_HOST = "controller"
```

允许所有主机访问仪表板：

```
 30 ALLOWED_HOSTS = ['*',]
```

配置 memcached 会话存储服务：

```
129 SESSION_ENGINE = 'django.contrib.sessions.backends.cache'
130 CACHES = {
131     'default': {
132         'BACKEND': 'django.core.cache.backends.memcached.MemcachedCache',
133         'LOCATION': 'controller:11211',
134     },
135 }
```

将其他的会话存储服务配置注释掉：

```
137 #CACHES = {
```

```
138 #     'default': {
139 #         'BACKEND': 'django.core.cache.backends.locmem.LocMemCache',
140 #     },
141 #}
```

启用第 3 版认证 API：

```
161 OPENSTACK_KEYSTONE_URL = "http://%s:5000/v3" % OPENSTACK_HOST
```

启用对域的支持：

```
64 OPENSTACK_KEYSTONE_MULTIDOMAIN_SUPPORT = True
```

配置 API 版本：

```
55 OPENSTACK_API_VERSIONS = {
56     "identity": 3,
57     "volume": 2,
58     "compute": 2,
59 }
```

通过仪表盘创建用户时的默认域配置为 default：

```
71 OPENSTACK_KEYSTONE_DEFAULT_DOMAIN = 'default'
```

通过仪表盘创建的用户默认角色配置为 user：

```
161 OPENSTACK_KEYSTONE_DEFAULT_ROLE = "user"
```

如果选择网络参数 1，禁用支持 3 层网络服务：

```
261 OPENSTACK_NEUTRON_NETWORK = {
262     'enable_router': False,
263     'enable_quotas': False,
265     'enable_distributed_router': False,
266     'enable_ha_router': False,
267     'enable_lb': False,
268     'enable_firewall': False,
269     'enable_vpn': False,
270     'enable_fip_topology_check': False,
```

可以选择性地配置时区：

```
371 TIME_ZONE = "Asia/Shanghai"
```

重启 Web 服务器以及会话存储服务：

```
[root@controller ~]# systemctl restart httpd.service memcached.service
```

2．验证操作

在浏览器中输入 http://controller/dashboard 访问仪表盘。

验证使用 admin 或者 demo 用户凭证和 default 域凭证登录，如图 2.1 所示。

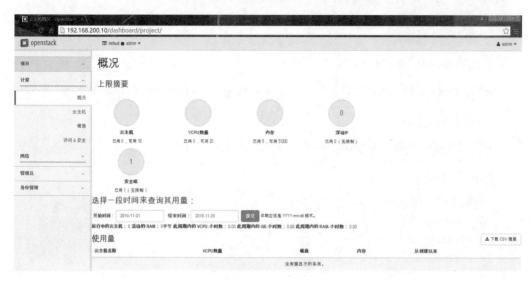

图 2.1　Dashboard

安装和配置好仪表板后，可以完成以下任务：

给用户提供公共 IP 地址、用户名和密码，这样就可以通过 Web 浏览器访问控制面板。在遇到任何 SSL 认证连接问题的情况下，指向服务 IP 到一个域名，让用户访问。

2.8　启动一个实例

这部分创建了必要的虚拟网络来支持创建实例。网络选项 1 包含一个使用公共虚拟网络（外部网络）的实例。网络选项 2 包含一个使用公共虚拟网络的实例、一个使用私有虚拟网络（私有网络）的实例。这部分教程在控制节点上使用命令行（CLI）工具。CLI 工具的更多信息，参考 OpenStack 用户手册。使用控制台，参考 OpenStack 用户手册 <http://docs.openstack.org/user-guide/dashboard.html>。

1. 创建虚拟网络

根据在网络选项中的选择来创建虚拟网络。如果选择选项 1，只需创建一个公有网络。如果选择选项 2，同时创建一个公有网络和一个私有网络。

2. 创建提供者网络

在启动实例之前，必须创建必须的虚拟机网络设施。对于网络选项 1，实例使用提供者（外部）网络，提供者网络通过 L2（桥/交换机）设备连接到物理网络。这个网络包括为实例提供 IP 地址的 DHCP 服务器。

admin 或者其他权限用户必须创建这个网络，因为它直接连接到物理网络设施。

（1）在控制节点上，加载 admin 凭证来获取管理员能执行的命令访问权限：

[root@controller ~]# . admin-openrc

（2）创建网络：

[root@controller ~]# neutron net-create --shared --provider:physical_network provider
　　--provider:network_type flat provider
Created a new network:

Field	Value
admin_state_up	True
availability_zone_hints	
availability_zones	
created_at	2016-11-29T11:26:41
description	
id	1a40813e-7bbd-4490-8f45-53bc9bb6fadb
ipv4_address_scope	
ipv6_address_scope	
mtu	1500
name	provider
port_security_enabled	True
provider:network_type	flat
provider:physical_network	provider
provider:segmentation_id	
router:external	False
shared	True
status	ACTIVE
subnets	
tags	
tenant_id	524aa64844814bc5bf6f73fd9e416410
updated_at	2016-11-29T11:26:43

--shared 选项允许所有项目使用虚拟网络。

[root@controller ~]# vim /etc/neutron/plugins/ml2/ml2_conf.ini
[ml2_type_flat]
153 flat_networks = provider

[root@controller ~]# vim /etc/neutron/plugins/ml2/linuxbridge_agent.ini
[linux_bridge]
138 physical_interface_mappings = provider:eno33554952

（3）在网络上创建一个子网

```
[root@controller ~]# neutron subnet-create --name provider --allocation-pool start=192.168.200.80,
end=192.168.200.100 --dns-nameserver 202.106.0.20 --gateway 192.168.200.1 provider
192.168.200.0/24
Created a new subnet:
+-------------------+----------------------------------------------------+
| Field             | Value                                              |
+-------------------+----------------------------------------------------+
| allocation_pools  | {"start": "192.168.200.80", "end": "192.168.200.100"} |
| cidr              | 192.168.200.0/24                                   |
| created_at        | 2016-11-29T11:32:44                                |
| description       |                                                    |
| dns_nameservers   | 202.106.0.20                                       |
| enable_dhcp       | True                                               |
| gateway_ip        | 192.168.200.100                                    |
| host_routes       |                                                    |
| id                | 63f4a7ef-7447-498e-ab95-546f1c2e0f58               |
| ip_version        | 4                                                  |
| ipv6_address_mode |                                                    |
| ipv6_ra_mode      |                                                    |
| name              | provider                                           |
| network_id        | 1a40813e-7bbd-4490-8f45-53bc9bb6fadb               |
| subnetpool_id     |                                                    |
| tenant_id         | 524aa64844814bc5bf6f73fd9e416410                   |
| updated_at        | 2016-11-29T11:32:44                                |
+-------------------+----------------------------------------------------+
```

使用提供者物理网络的子网 CIDR 标记替换 PROVIDER_NETWORK_CIDR。

将 START_IP_ADDRESS 和 END_IP_ADDRESS 使用想分配给实例的子网网段的第一个和最后一个 IP 地址。这个范围不能包括任何已经使用的 IP 地址。

将 DNS_RESOLVER 替换为 DNS 解析服务的 IP 地址。在大多数情况下，可以从主机 /etc/resolv.conf 文件选择一个使用。

将 PUBLIC_NETWORK_GATEWAY 替换为公共网络的网关，一般的网关 IP 地址以 ".1" 结尾。

3. 创建 m1.nano 规格的主机

默认的最小规格的主机需要 512 MB 内存。对于环境中计算节点内存不足 4 GB 的，推荐创建只需要 64 MB 的 m1.nano 规格的主机。若单纯为了测试的目的，请使用 m1.nano 规格的主机来加载 CirrOS 镜像。

```
[root@controller ~]# openstack flavor create --id 0 --vcpus 1 --ram 64 --disk 1 m1.nano
+----------------------------+---------+
| Field                      | Value   |
```

```
+----------------------------+---------+
| OS-FLV-DISABLED:disabled   | False   |
| OS-FLV-EXT-DATA:ephemeral  | 0       |
| disk                       | 1       |
| id                         | 0       |
| name                       | m1.nano |
| os-flavor-access:is_public | True    |
| ram                        | 64      |
| rxtx_factor                | 1.0     |
| swap                       |         |
| vcpus                      | 1       |
+----------------------------+---------+
```

4. 生成一个键值对

大部分云镜像支持公共密钥认证而不是传统的密码认证。在启动实例前，必须添加一个公共密钥到计算服务。

（1）导入租户 demo 的凭证

```
[root@controller ~]# . demo-openrc
```

（2）生成和添加密钥对

```
[root@controller ~]# ssh-keygen -q -N ""
Enter file in which to save the key (/root/.ssh/id_rsa):
[root@controller ~]# openstack keypair create --public-key ~/.ssh/id_rsa.pub mykey
+-------------+-------------------------------------------------+
| Field       | Value                                           |
+-------------+-------------------------------------------------+
| fingerprint | b5:82:c3:1d:16:b1:6a:d1:ed:d3:ae:61:ed:75:d2:f4 |
| name        | mykey                                           |
| user_id     | c7552cc0ee694bc7b5a855cbbd4165a5                |
+-------------+-------------------------------------------------+
```

（3）验证公钥的添加

```
[root@controller ~]# openstack keypair list
+-------+-------------------------------------------------+
| Name  | Fingerprint                                     |
+-------+-------------------------------------------------+
| mykey | b5:82:c3:1d:16:b1:6a:d1:ed:d3:ae:61:ed:75:d2:f4 |
+-------+-------------------------------------------------+
```

5. 增加安全组规则

默认情况下，default 安全组适用于所有实例并且包括拒绝远程访问实例的防火墙规则。对诸如 CirrOS 这样的 Linux 镜像，推荐至少允许 ICMP（ping）和安全 shell（SSH）规则。

添加规则到 default 安全组。

（1）允许 ICMP（ping）：

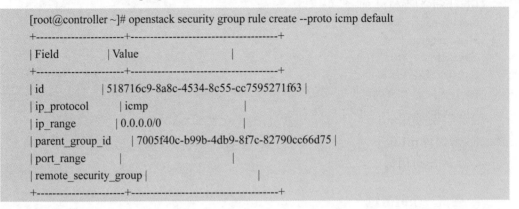

```
[root@controller ~]# openstack security group rule create --proto icmp default
+-------------------+--------------------------------------+
| Field             | Value                                |
+-------------------+--------------------------------------+
| id                | 518716c9-8a8c-4534-8c55-cc7595271f63 |
| ip_protocol       | icmp                                 |
| ip_range          | 0.0.0.0/0                            |
| parent_group_id   | 7005f40c-b99b-4db9-8f7c-82790cc66d75 |
| port_range        |                                      |
| remote_security_group |                                  |
+-------------------+--------------------------------------+
```

（2）允许安全 shell（SSH）的访问：

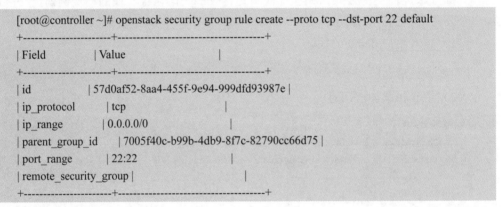

```
[root@controller ~]# openstack security group rule create --proto tcp --dst-port 22 default
+-------------------+--------------------------------------+
| Field             | Value                                |
+-------------------+--------------------------------------+
| id                | 57d0af52-8aa4-455f-9e94-999dfd93987e |
| ip_protocol       | tcp                                  |
| ip_range          | 0.0.0.0/0                            |
| parent_group_id   | 7005f40c-b99b-4db9-8f7c-82790cc66d75 |
| port_range        | 22:22                                |
| remote_security_group |                                  |
+-------------------+--------------------------------------+
```

6. 启动一个实例

如果选择网络选项 1，只能在公网创建实例。如果选择网络选项 2，可以在公网或私网创建实例。

在公有网络上创建实例步骤如下。

（1）确定实例选项：

启动一台实例，您必须至少指定一个类型、镜像名称、网络、安全组、密钥和实例名称。

（2）在控制节点上，获得 admin 凭证来获取只有管理员能执行的命令的访问权限：

```
[root@controller ~]# . demo-openrc
```

（3）一个实例指定了虚拟机资源的大致分配，包括处理器、内存和存储。

列出可用类型：

```
[root@controller ~]# openstack flavor list
+----+-----------+-------+------+-----------+-------+-----------+
| ID | Name      | RAM   | Disk | Ephemeral | VCPUs | Is Public |
```

```
+----+-----------+--------+-------+-----------+--------+-----------+
| 0  | m1.nano   | 64     | 1     |       0   | 1      | True      |
| 1  | m1.tiny   | 512    | 1     |       0   | 1      | True      |
| 2  | m1.small  | 2048   | 20    |       0   | 1      | True      |
| 3  | m1.medium | 4096   | 40    |       0   | 2      | True      |
| 4  | m1.large  | 8192   | 80    |       0   | 4      | True      |
| 5  | m1.xlarge | 16384  | 160   |       0   | 8      | True      |
+----+-----------+--------+-------+-----------+--------+-----------+
```

这个实例使用 m1.tiny 规格的主机。如果创建了 m1.nano 这种主机规格，使用 m1.nano 来代替 m1.tiny。

（4）列出可用镜像：

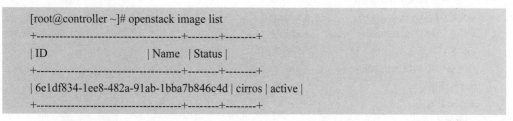

这个实例使用 cirros 镜像。

（5）列出可用网络：

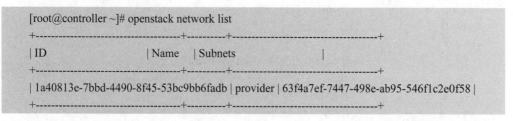

这个实例使用 provider 公有网络。必须使用 ID 而不是名称才可以使用这个网络。

（6）列出可用的安全组：

```
[root@controller ~]# openstack security group list
+--------------------------------------+---------+------------------------+---------+
| ID                                   | Name    | Description            | Project |
+--------------------------------------+---------+------------------------+---------+
| 7005f40c-b99b-4db9-8f7c-82790cc66d75 | default | Default security group |
524aa64844814bc5bf6f73fd9e416410 |
+--------------------------------------+---------+------------------------+---------+
```

这个实例使用 default 安全组。

7. 创建实例

（1）启动实例：

如果选择选项 1 并且你的环境只有一个网络，可以省去"-nic"选项因为 OpenStack 会自动选择这个唯一可用的网络。

使用 provider 公有网络的 ID 替换 PUBLIC_NET_ID。

```
[root@controller ~]# openstack server create --flavor m1.nano --image cirros --nic net-id=1a40813e-7bbd-4490-8f45-53bc9bb6fadb --security-group default --key-name mykey provider-instance
+--------------------------------------+-------------------------------------------------+
| Field                                | Value                                           |
+--------------------------------------+-------------------------------------------------+
| OS-DCF:diskConfig                    | MANUAL                                          |
| OS-EXT-AZ:availability_zone          |                                                 |
| OS-EXT-SRV-ATTR:host                 | None                                            |
| OS-EXT-SRV-ATTR:hypervisor_hostname  | None                                            |
| OS-EXT-SRV-ATTR:instance_name        | instance-00000006                               |
| OS-EXT-STS:power_state               | 0                                               |
| OS-EXT-STS:task_state                | scheduling                                      |
| OS-EXT-STS:vm_state                  | building                                        |
| OS-SRV-USG:launched_at               | None                                            |
| OS-SRV-USG:terminated_at             | None                                            |
| accessIPv4                           |                                                 |
| accessIPv6                           |                                                 |
| addresses                            |                                                 |
| adminPass                            | a754mQTwUPSM                                    |
| config_drive                         |                                                 |
| created                              | 2016-11-29T13:09:01Z                            |
| flavor                               | m1.tiny (1)                                     |
| hostId                               |                                                 |
| id                                   | 8dc98c38-09f5-4248-8e4d-fd4bfaedf7c3            |
| image                                | cirros (6e1df834-1ee8-482a-91ab-1bba7b846c4d)   |
| key_name                             | mykey                                           |
| name                                 | provider-instance                               |
| os-extended-volumes:volumes_attached | []                                              |
| progress                             | 0                                               |
| project_id                           | 524aa64844814bc5bf6f73fd9e416410                |
| properties                           |                                                 |
| security_groups                      | [{u'name': u'default'}]                         |
| status                               | BUILD                                           |
| updated                              | 2016-11-29T13:09:01Z                            |
| user_id                              | c7552cc0ee694bc7b5a855cbbd4165a5                |
+--------------------------------------+-------------------------------------------------+
```

（2）检查实例的状态：

```
[root@controller ~]# openstack server list
+--------------------------------------+-------------------+--------+------------------------+
| ID                                   | Name              | Status | Networks               |
+--------------------------------------+-------------------+--------+------------------------+
| 8dc98c38-09f5-4248-8e4d-fd4bfaedf7c3 | provider-instance | BUILD  | provider=192.168.200.83|
+--------------------------------------+-------------------+--------+------------------------+
```

当构建过程完全成功后，状态会从 BUILD 变为 ACTIVE。

8. 使用虚拟控制台访问实例

获取实例的 Virtual Network Computing（VNC）会话 URL 并从 Web 浏览器访问它：

```
[root@controller ~]# openstack console url show provider-instance
+-------+-----------------------------------------------------------------+
| Field | Value                                                           |
+-------+-----------------------------------------------------------------+
| type  | novnc                                                           |
| url   | http://controller:6080/vnc_auto.html?token=05b62daf-ef91-4aef-a872-19918130de60 |
+-------+-----------------------------------------------------------------+
```

验证能否远程访问实例。

本章总结

- 手工部署 OpenStack 的环境准备，需要 3 台主机，然后分别进行基础配置。
- OpenStack:Identity service 为认证管理，授权管理和服务目录服务管理提供单点整合。
- 镜像服务（glance）允许用户发现、注册和获取虚拟机镜像。
- OpenStack 计算组件请求 OpenStack Identity 服务进行认证；请求 OpenStack Image 服务提供磁盘镜像；为 OpenStack dashboard 提供用户与管理员接口。
- OpenStack Networking（neutron），允许创建、插入接口设备，这些设备由其他的 OpenStack 服务管理。插件式的实现可以容纳不同的网络设备和软件，为 OpenStack 架构与部署提供了灵活性。
- Dashboard（horizon）是一个 Web 接口，使得云平台管理员以及用户可以管理不同的 OpenStack 资源以及服务。

随手笔记

第 3 章

云存储

技能目标

- 理解对象存储与块存储的概念
- 理解 swift 的核心概念
- 理解 cinder 的核心概念

本章导读

市面上常见的主流的存储类型包括块存储与文件存储，DAS 和 SAN 都是典型的块存储类型，NAS 是文件级存储。

对象存储（Object-based Storage）是一种新的网络存储架构，兼具 SAN 高速直接访问磁盘特点及 NAS 的分布式共享特点。块存储服务（Cinder）提供块存储，块存储适合性能敏感性业务场景，例如数据库存储。

知识服务

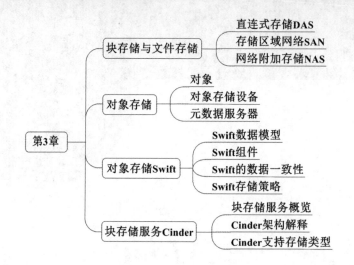

3.1 块存储与文件存储

市面上常见的主流的存储类型包括块存储与文件存储。而目前常见的三种存储方式 DAS、SAN、NAS 中 DAS 和 SAN 都是典型的块存储类型，NAS 是文件级存储，均被广泛应用于企业存储设备中。

1. 直连式存储 DAS

DAS（Direct Attach Storage）是直接连接于主机服务器的一种储存方式，每一台主机服务器有独立的存储设备，每台主机服务器的存储设备无法互通，通常用在单一网络且数据交换量不大，性能要求不高的环境下。

2. 存储区域网络 SAN

SAN（Storage Area Network）是一种用高速（光纤）网络连接专业主机服务器的一种存储方式，此系统位于主机群的后端，它使用高速 I/O 连接方式，如 SCSI、Fibre Channel。采用 SCSI 块 I/O 的命令集，通过在磁盘或 FC（Fiber Channel）级的数据访问提供高性能的随机 I/O 和数据吞吐率，具有高带宽、低延迟的优势。一般而言，SAN 应用在对网络速度要求高、对数据的可靠性和安全性要求高、对数据共享的性能要求高的应用环境中，在高性能计算中占有一席之地，但是由于 SAN 系统的价格较高，且可扩展性较差，已不能满足成千上万个 CPU 规模的系统。

3. 网络附加存储 NAS

NAS（Network Attached Storage）：是一套网络存储设备，通常是直接连在网络上并提供资料存取服务，一套 NAS 存储设备就如同一个提供数据文件服务的系统。它采用 NFS 或 CIFS 命令集访问数据，以文件为传输协议，通过 TCP/IP 实现网络化存储，可扩展性好、用户易管理、性价比高，常用在诸如教育、政府、企业等数据存储应用。

目前在集群计算中应用较多的也是 NFS 文件系统，但由于 NAS 的协议开销高、带宽低、延迟大，不利于在高性能集群中应用。

针对 Linux 集群对存储系统高性能和数据共享的需求，国际上已开始研究全新的存储架构和新型文件系统，希望能有效结合 SAN 和 NAS 系统的优点，支持直接访问磁盘以提高性能，通过共享的文件和元数据以简化管理。目前对象存储系统已成为 Linux 集群系统中高性能存储系统的研究热点。

3.2 对象存储

对象存储（Object-based Storage）是一种新的网络存储架构。兼具 SAN 高速直接访问磁盘的特点及 NAS 分布式共享的特点。其核心是将数据通路（数据读或写）和控制通路（元数据）分离，并且基于对象存储设备（Object-based Storage Device，OSD）构建存储系统，每个对象存储设备具有一定的智能，能够自动管理其上的数据分布。目前国际上通常采用刀片式结构实现对象存储设备。

3.2.1 对象存储结构组成

对象存储结构组成包括：对象、对象存储设备、元数据服务器、对象存储系统的客户端几个组成部分。

1. 对象

对象是系统中数据存储的基本单位，一个对象实际上就是文件的数据和一组属性信息（Meta Data）的组合。

2. 对象存储设备

对象存储设备也叫做 OSD（Object-based Storage Device），是一个智能设备，具有自己的 CPU、内存、网络和磁盘系统。它同块设备的区别在于两者提供的访问接口不同。OSD 的主要功能包括数据存储和安全访问。尤其是以下这三个主要功能：

（1）数据存储

OSD 管理对象数据，并将它们放置在标准的磁盘系统上，OSD 并不提供块接口访问方式，而是由 Client 请求数据时用对象 ID、偏移进行数据读写。

（2）智能分布

OSD 用其自身的 CPU 和内存优化数据分布，支持数据的预取。由于 OSD 可以智能地支持对象的预取从而可以优化磁盘的性能。

（3）每个对象元数据的管理

OSD 管理存储在其上对象的元数据，该元数据与传统的 inode 元数据相似，通常包括对象的数据块和对象的长度。而在传统的 NAS 系统中，这些元数据是由文件服务器进行维护。对象存储架构将系统中主要的元数据管理工作由 OSD 来完成，同时也降低了 Client 的开销。

3. 元数据服务器

元数据服务器（Metadata Server，MDS）：控制 Client 与 OSD 对象的交互，主要提供以下几个功能：

(1) 对象存储访问

MDS 构造、管理描述每个文件分布的视图，为 Client 提供访问该文件所含对象的能力，并且允许 Client 直接访问对象。

OSD 在接收到每个请求时将先验证该能力，然后才可以进行访问。

(2) 文件和目录访问管理

MDS 在存储系统上构建一个文件结构，包括限额控制、目录和文件的创建和删除、访问控制等。

(3) Client Cache 一致性

为了提高 Client 性能，在对象存储系统设计时通常支持 Client 方的 Cache。由于引入 Client 方的 Cache，带来了 Cache 一致性问题，当 Cache 的文件发生改变时，将通知 Client 刷新 Cache，从而防止 Cache 不一致引发的问题。

(4) 对象存储系统的客户端 Client

为了有效支持 Client 访问 OSD 上的对象，需要在计算节点实现对象存储系统的 Client，通常提供 POSIX 文件系统接口，允许应用程序像执行标准的文件系统操作一样。

3.3 对象存储 Swift

Swift 是用来创建可扩展的、冗余的、开源的对象存储（引擎），可以使用标准化的服务器存储 PB 级可用数据。Swift 不是文件系统也不是数据库，而是使用 account-container-object 概念存储 object，适合存储非结构化数据存储，如虚拟机镜像、图片存储、邮件存储、文档备份等。

3.3.1 Swift 数据模型

Swift 采用层次数据模型，共设三层逻辑结构：Account/Container/Object（即账户/容器/对象），每层节点数均没有限制，可以任意扩展，如图 3.1 所示。

具体来说 Account 是出于访问安全性考虑，使用 Swift 系统的每个用户必须有一个账号（Account）。只有通过 Swift 验证的账号才能访问 Swift 系统中的数据。提供账号验证的节点被称为 Account Server。可由 Swauth 提供账号权限认证服务。用户通过账号验证后将获得一个验证字符串（authentication token），后续的每次数据访问操作都需要传递这个字符串。Container 是 Swift 中类似于 Windows 操作系统中的文件夹或者 UNIX 类操作系统中的目录，用于组织管理数据，所不同的是 Container 不能嵌套。数据都以 Object 的形式存放在 Container 中。

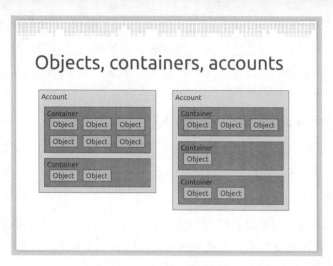

图 3.1 Swift 逻辑结构

简单来说就是 Account 对应租户，用于隔离；Container 对应某个租户数据的存储区域；Object 对应存储区域中具体的 Block。

Swift 存储数据的过程类似文件存储过程，如图 3.2 所示。

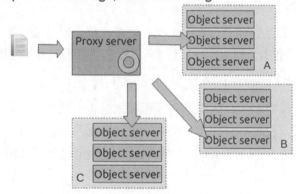

图 3.2 Swift 存储数据过程

比如，我们存储一个文件，会在对应的磁盘块组中对应的元数据区域产生对应的 inode，基于这个 inode 将数据存储到数据区中对应的块组中的 Block。

3.3.2 Swift 组件

Swift 并不是单一的存在，它包含很多组件，下面就来具体了解下 Swift 所包含的这些组件。

1. 代理服务（Proxy Service）

代理服务对外提供对象服务 API，会根据环的信息来查找服务地址并转发用户请求至相应的账户、容器或者对象服务；由于采用无状态的 REST 请求协议，可以进行横向扩展来进行均衡负载。

2. 认证服务（Authentication Service）

认证服务用来验证访问用户的身份信息，并获得一个在一定的时间内会一直有效的对象访问令牌（Token），认证服务会验证访问令牌的有效性并缓存下来直至过期时间。

3. 缓存服务（Cache Service）

缓存服务用于内容的缓存，缓存包括对象服务令牌，账户和容器的存在信息，但不会对象本身的数据进行缓存。缓存服务可采用 Memcached 集群来实现，Swift 会使用一致性散列算法来分配 Memcached 缓存地址。

4. 账户服务（Account Service）

账户服务用于提供账户元数据和统计信息，并维护所含容器列表的服务，将每个账户的信息存储在一个 SQLite 数据库中。

5. 容器服务（Container Service）

容器服务用来提供容器元数据和统计信息，同时对所含对象列表的服务进行维护，每个容器的信息也存储在一个 SQLite 数据库中。

6. 对象服务（Object Service）

对象服务用来提供对象元数据和内容服务，每个对象的内容会以文件的形式存储在文件系统中，元数据会作为文件属性来存储，建议采用支持扩展属性的 XFS 文件系统。

7. 复制服务（Replicator）

复制服务会检测本地分区副本和远程副本是否一致，具体是通过对比散列文件和高级水印来完成，发现不一致时会采用推式（Push）更新远程副本，例如对象复制服务会使用远程文件复制工具 rsync 来同步。另外一个任务是确保被标记删除的对象从文件系统中移除。

8. 更新服务（Updater）

更新服务可以当对象由于高负载的原因而无法立即更新时，将任务序列化到本地文件系统中进行排队，以便服务恢复后进行异步更新。例如成功创建对象后容器服务器没有及时更新对象列表，这个时候容器的更新操作就会进入排队中，更新服务会在系统恢复正常后扫描队列并进行相应的更新处理。

9. 审计服务（Auditor）

审计服务用来检查对象、容器和账户的完整性，如果发现比特级的错误，文件将被隔离，并复制其他的副本以覆盖本地损坏的副本；其他类型的错误会被记录到日志中。

10. 账户清理服务（Account Reaper）

账户清理服务用来移除被标记为删除的账户，删除其所包含的所有容器和对象。

3.3.3 Swift 的数据一致性

存储在 Swift 里面的数据有好几个备份，而且各个节点之间是平等的关系，没有"主节点"这个概念，因此任意一个节点出现故障时，数据并不会丢失。Swift 必须面对的一个问题就是如何保持数据的一致性。因为一个文件并不是只保存一份的，在 Swift 中默认要保存 3 个副本，当更新的时候这 3 个文件要同时更新，当其中一个文件损坏时必须能迅速地复制一份完整的文件来替换。

为了保证 Swift 数据的一致性，Swift 有 3 个服务来解决这个问题：Auditor、Updater 和 Replicator。Auditor 运行在每个 Swift 服务器的后台持续地扫描磁盘来检测数据的完整性。如果发现数据损坏，Auditor 就会将该文件移动到隔离区域，然后由 Replicator 负责用一个完好的副本文件来替代该数据。如果更新失败，该次更新在本地文件系统上会被加入队列，然后 Updaters 会继续处理这些失败了的更新工作。

3.3.4 Swift 存储策略

在对象存储中存储的不仅是数据还有丰富的与数据相关的属性信息。系统会给每一个对象分配一个唯一的 ID。对象本身是平等的，所有的 ID 都属于一个平坦的地址空间，而并非文件系统那样的树状逻辑结构，如图 3.3 所示。这种存储结构带来的好处是可以实现数据的智能化管理，因为对象本身包含了元数据信息，甚至更多的属性，我们可以根据这些信息对对象进行高效的管理。

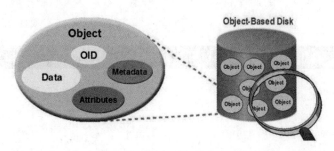

图 3.3 Swift 存储策略

那么 Client 要存储一个文件，用什么策略决定存在哪台 Storage Node 上呢？例如我们可以制定这样的存储策略，如果对象中包含 priority:high 这样的属性，我们就对文件做比平常文件多的备份次数。平坦地址空间的设计使得访问对象只通过一个唯一的 ID 标识即可，不需要复杂的路径结构。因此可以知道 Swift 中没有"路径"这个概念，所以也没有所谓的"文件夹"这样的概念。

那么如何根据对象 ID 把对象存在合适的 Storage Node 呢？

解决的方法是使用"一致性 Hash 法"。一致性哈希算法的基本实现原理是将机器节点和 Key（在本文里就是对象的 ID）值都按照一样的 hash 算法映射到一个 $0 \sim 2^{32}$ 的圆环上。当有一个写入缓存的请求到来时，计算 Key 值 K 对应的哈希值 Hash(K)，如果该值正好对应之前某个机器节点的 Hash 值，则直接写入该机器节点，如果没有对应的机器节点，则顺时针查找下一个节点，进行写入，如果超过 2^{32} 还没找到对应节点，则从 0 开始查找（因为是环状结构），如图 3.4 所示。

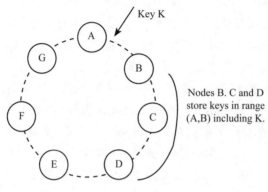

图 3.4　一致性 Hash 法

经过一致性哈希算法散列之后，当有新的机器加入时，将只影响一台机器的存储情况，例如新加入的节点 H 的散列在 B 与 C 之间，则原先由 C 处理的一些数据可能将移至 H 处理，而其他所有节点的处理情况都将保持不变。而如果删除一台机器，例如删除 C 节点，此时原来由 C 处理的数据将移至 D 节点，而其他节点的处理情况仍然不变。而由于在机器节点散列和缓冲内容散列时都采用了同一种散列算法，因此也很好地降低了分散性和负载。

现将所有 Swift 支持的操作总结如表 3-1 所示。

表 3-1　Swift RESTful API 总结

资源类型	URL	GET	PUT	POST	DELETE	HEAD
账户	/account/	获取容器列表				获取账户元数据
容器	/account/container	获取对象列表	创建容器	更新容器元数据	删除容器	获取容器元数据
对象	/account/container/object	获取对象内容和元数据	创建、更新或复制对象	更新对象元数据	删除对象	获取对象元数据

3.3.5　对象存储 Swift 补充

1. 副本

如果集群中的数据在本地节点上只有一份，一旦发生故障就可能会造成数据的永久性丢失。因此，需要有冗余的副本来保证数据安全。

Swift 中引入了副本的概念，其默认值为 3，理论依据主要来源于 NWR 策略（也叫 Quorum 协议）。NWR 是一种在分布式存储系统中用于控制一致性级别的策略。

在 Amazon 的 Dynamo 云存储系统中，使用了 NWR 来控制一致性。其中，N 代表同一份数据的副本的份数，W 是更新一个数据对象时需要确保成功更新的份数；R 代表读取一个数据需要读取的副本的份数。公式 W+R>N，保证某个数据不被两个不同的事务同时读和写；公式 W>N/2，保证两个事务不能并发写某一个数据。

在分布式系统中，数据的单点是不允许存在的。即线上正常存在的副本数量为 1 的情况是非常危险的，因为一旦这个副本再次出错，就可能发生数据的永久性错误。假如我们把 N 设置成为 2，那么只要有一个存储节点发生损坏，就会有单点的存在，所以 N 必须大于 2。N 越高，系统的维护成本和整体成本就越高。工业界通常把 N 设置为 3。例如，对于 MySQL 主从结构，其 NWR 数值分别是 N= 2, W = 1, R = 1，没有满足 NWR 策略。而 Swift 的 N=3, W=2, R=2，完全符合 NWR 策略，因此 Swift 系统是可靠的，没有单点故障。

2．Zone

如果所有的 Node 都在一个机架或一个机房中，那么一旦发生断电、网络故障等，都将造成用户无法访问。因此需要一种机制对机器的物理位置进行隔离，以满足分区容忍性（CAP 理论中的 P）。

因此，引入了 Zone 的概念，把集群的 Node 分配到每个 Zone 中。其中同一个 Partition 的副本不能同时放在同一个 Node 上或同一个 Zone 内，如图 3.5 所示。注意，Zone 的大小可以根据业务需求和硬件条件自定义，可以是一块磁盘、一台存储服务器，也可以是一个机架甚至一个 IDC。

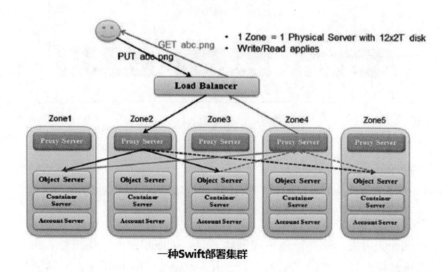

图 3.5　Zone 概念的引入

3.4 块存储服务 Cinder

块存储服务（cinder）提供块存储。存储的分配和消耗是由块存储驱动器，或者多后端配置的驱动器决定的。还有很多驱动程序可用：NAS/SAN，NFS，ISCSI，Ceph 等。块存储适合性能敏感性业务场景，例如数据库存储大规模可扩展的文件系统或服务器需要访问到块级的裸设备存储。

典型情况下，块服务 API 和调度器服务运行在控制节点上。取决于使用的驱动，卷服务器可以运行在控制节点、计算节点或单独的存储节点。

3.4.1 块存储服务概览

块存储服务为 OpenStack 中的实例提供持久的存储，块存储提供一个基础设施用于管理卷，以及和 OpenStack 计算服务交互，为实例提供卷。此服务也会激活管理卷的快照和卷类型的功能。

3.4.2 块存储服务组件

块存储服务也包含很多组件，各组件之间的关系如图 3.6 所示。

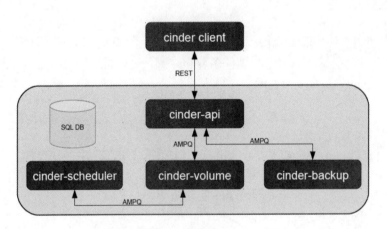

图 3.6 Cinder 组件

1. cinder-api

cinder-api 用来接受 API 请求，并将其路由到 cinder-volume 执行。

2. cinder-volume

cinder-volume 用来与块存储服务和例如 cinder-scheduler 的进程进行直接交互。它也可以与这些进程通过一个消息队列进行交互。cinder-volume 服务响应送到块存储服务的读写请求来维持状态。它也可以和多种存储提供者在驱动架构下进行交互。

3. cinder-scheduler 守护进程

cinder-scheduler 守护进程会选择最优存储提供节点来创建卷。其与 nova-scheduler 组件类似。

4. cinder-backup 守护进程

cinder-backup 服务提供任何种类备份卷到一个备份存储提供者。就像 cinder-volume 服务，它与多种存储提供者在驱动架构下进行交互。

5. 消息队列

消息队列作用是在块存储的进程之间路由信息。

3.4.3 Cinder 架构解释

Cinder 主要核心是对卷的管理，允许对卷、卷的类型、卷的快照进行处理。它并没有实现对块设备的管理和实际服务，而是为后端不同的存储结构提供了统一的接口，不同的块设备服务厂商在 Cinder 中实现其驱动支持以与 OpenStack 进行整合。在 CinderSupportMatrix 中可以看到众多存储厂商如 HP、NetAPP、IBM、SolidFire、EMC 和众多开源块存储系统对 Cinder 的支持。其核心架构由 API Service、Scheduler Service、Volume Service 三部分组成，它们之间的关系如图 3.7 所示。

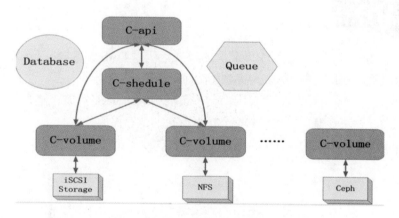

图 3.7 Cinder 核心架构

1. API Service

API Service 负责接受和处理请求，并将请求放入 RabbitMQ 队列。

2. Scheduler Service

Scheduler Service 处理任务队列的任务，并根据预定策略选择合适的 Volume Service 节点来执行任务。目前版本的 Cinder 仅仅提供了一个 Simple Scheduler，该调度器选择卷数量最少的一个活跃节点来创建卷。

3. Volume Service

Volume Service 该服务运行在存储节点上，管理存储空间。每个存储节点都有一个 Volume Service，若干个这样的存储节点联合起来可以构成一个存储资源池。为了支持不同类型和型号的存储，均通过 Drivers 的形式为 Cinder 的 Volume Service 提供相应的 Cinder-Volume。

3.4.4 Cinder 支持存储类型

Cinder 目前支持的存储类型有：
本地存储如 LVM，Sheepdog；
网络存储如 NFS，Ceph；
HP 的 3PAR(iSCSI/FC)，LeftHand(iSCSI) 存储设备；
IBM 的 Storwize family/SVC (iSCSI/FC)，XIV (iSCSI)，GPFS，zVM 设备；
Netapp 的 NetApp(iSCSI/NFS) 设备；
EMC 的 VMAX/VNX (iSCSI)，Isilon(iSCSI) 设备；
Solidfire 的 Solidfire cluster(iSCSI) 设备。
关于 Cinder 的配置请上课工场 APP 或官网 kgc.cn 观看视频。

本章总结

- 所有磁盘阵列都是基于 Block 块的模式，所以 DAS 和 SAN 都是典型的块存储类型，而所有的 NAS 产品都是文件级存储。
- Swift 是开源的对象存储（引擎），提供给 OpenStack 对象存储服务。
- Cinder 是块存储，提供给 OpenStack 永久的块存储服务。
- Swift 和 Cinder 都属于开源项目，由各自的核心组件组成。

第4章

Fuel 安装 OpenStack

技能目标

- 理解 Openstack 核心概念
- 会进行 Fuel 环境的部署
- 会使用 Fuel 安装 OpenStack

本章导读

Fuel 是由 Mirantis 公司开发的开源 OpenStack 部署管理工具，也被称为 Mirantis OpenStack，支持安装 CentOS 和 Ubuntu 系统，通过 Fuel Web 界面可以简单、快速地部署 OpenStack 环境，还可以动态添加计算、存储节点。

知识服务

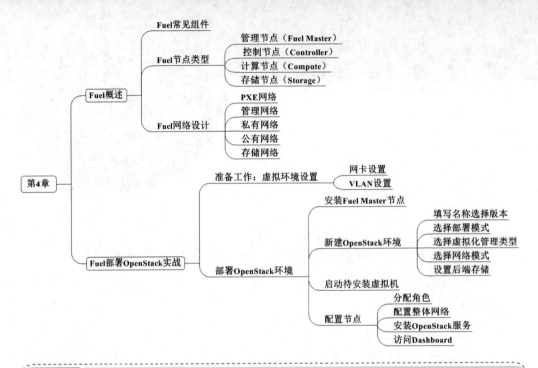

4.1 Fuel 概述

Fuel 是由 Mirantis 公司开发的开源 OpenStack 部署管理工具，也被称为 Mirantis OpenStack，支持安装 CentOS 和 Ubuntu 系统，通过 Fuel Web 界面可以简单、快速地部署 OpenStack 环境，还可以动态添加计算、存储节点。

1．Fuel 常见组件

Fuel 是由多种独立组件构成，有一些是 Fuel 自己定义的组件，也有一些是目前比较火的第三方软件，Fuel 的组件及功能如表 4-1 所示。

表 4-1　Fuel 常见组件

组件	功能
UI	基于 JavaScript 开发的前端以及骨干框架页面应用
Nailgun	基于 Python 开发用于部署和管理的 REST API，管理磁盘、网络相关数据，是 Fuel 的核心组件
Astute	通过 AMQP 与 Nailgun 进行数据交换，同时管理 Cobbler、Puppet、Shell 脚本等
Cobbler	提供快速从网络进行安装部署 Linux 系统及服务的功能
Puppet	提供自动化部署安装服务，通过 MCollective agent 管理其他的配置管理框架
Mcollective agents	执行类似硬件驱动清理、网络连接探查任务
OSTF	在 OpenStack 部署后用于测试环境，是独立的组件

Fuel 为支持不同硬件厂商提供了插件机制（如做监控的 Zabbix，提供 VPN 服务的 VPNaaS，提供防火墙服务的 FWaaS，分析日志的 ElasticSearch/Kibana），让用户自由选择启用插件。下面是 Mirantis 官方发布的插件，下载后直接在 Fuel 的 Master 里安装即可使用。

2. Fuel 节点类型

Fuel 部署 OpenStack 环境主要有四种节点类型：管理节点（Fuel Master）、控制节点（Controller）、计算节点（Compute）和存储节点（Storage）。其中整个部署是从管理节点上发起的。

Fuel Master 节点可以看做是安装 OpenStack 的工具集，包含了安装 OpenStack 组件、以及打包的 CentOS、Ubuntu 系统，以及提供 DHCP、TFTP、PXE、Cobber 裸机安装系统相关服务，以及 Puppet 等推送工具。

Controller 节点主要用于调度管理、API 服务、数据库管理、身份认证管理和镜像管理、网络管理等工作。

Compute 节点提供可使用的资源按需运行虚拟机，与 Controller 上的身份认证、镜像和网络进行交互，支持多种虚拟化。

需要注意的是 Fuel 不允许把控制节点和计算节点一起安装。

Storage 节点作为存储节点，可以独立部署，也可以直接在控制节点或者计算节点上运行存储服务。

3. Fuel 网络设计

Fuel 的安装成功与否关键在于网络的设置，如果网络验证不通过，就无法进一步安装 OpenStack。Fuel 中，将网络分为 PXE 网络、管理网络、公有网络、私有网络和存储网络。

PXE 网络，是 Fuel 安装各节点的网络，是节点在开机时设置的启动网络，该网络独立使用，并且不能开启 DHCP 服务。

管理网络，是整个 OpenStack 内部各个组件之间的通信使用的网络，比如与 Controller 节点进行交互的各 API 之间、身份认证服务、监控服务等。

私有网络，主要用于内部通信，划分 VLAN 供 OpenStack 租户使用，每一个子网都使用一个 VLAN 做隔离，确保不同租户都可以有自己的网络。

公有网络，主要用于外部访问，初次部署 Floating IP 也是属于这个网段的 IP 地址，部署完成后可以手动添加额外的 Floating IP 网段，使得外部用户可以通过 Floating IP 访问实例。

存储网络，负责专门给存储使用的私有网络。

本案例环境网络规划如表 4-2 所示。

表 4-2　网络规划

名称	网段	VLAN	备注
PXE	10.20.0.0/24	vlan 503	Fuel 安装节点网络
管理网络	192.168.0.0/24	vlan 501	
存储网络	192.168.1.0/24	vlan 502	
私有网络	IP 192.168.100.0/24 Gateway 192.168.100.1 DNS 8.8.4.4，8.8.8.8	vlan 510-530	用于虚拟机间通信
公有网络	172.16.200.0/24 浮动 IP 范围 172.16.200.130～254	vlan500	用于集群公网和虚拟主机浮动 IP，需要能与外网通信

Fuel 是针对生产环境下部署 OpenStack 环境，为了讲解安装过程在虚拟机上进行演示说明。VMware Workstation 虚拟网络设置概要如图 4.1 所示。需要注意的是必须取消 DHCP 服务，否则会影响之后 PXE 装机和网络验证。

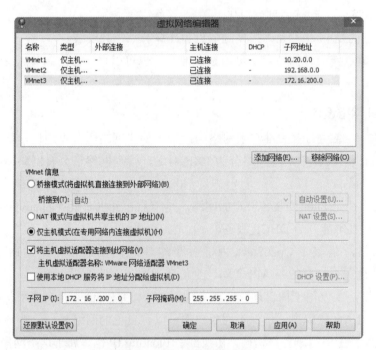

图 4.1　虚拟网络设置概要

4.2　虚拟环境设置

本案例使用四台 VMware Workstation 虚拟机，一台 Fuel Master 节点，一台 Controller 节点，一台 Compute 节点和一台 Storage 节点，均采用最小化安装。四个节点的内存至少 2GB，如果内存足够可以适当调整，硬盘容量至少 42GB 空间，如果小于 42GB 安装过程中会提示硬盘空间不足。

各节点网卡配置如表 4-3 所示。

表 4-3　网卡配置

主机名称	网卡配置	备注
Fuel Master	eth0 10.20.0.2	交换机 0～4 口
	eth1 172.16.200.2	交换机 5～10 口
Controller	eth0 10.20.0.0/24	交换机 0～4 口
Compute	eth1 公有网络 172.16.200.0/24 　　管理网络 192.168.0.0/24 　　存储网络 192.168.1.0/24	交换机 5～10 口
Storage	eth2 私有网络 192.168.100.0/24	交换机 11～15 口

1. 各节点设置

使用 VMware Workstation 虚拟机模拟硬件与软件设置，各节点设置如下，Fuel Master 节点设置概要如图 4.2 所示。

图 4.2　Fuel Master 节点设置

Controller 节点设置概要如图 4.3 所示。

图 4.3 Controller 节点设置

Compute 节点需要勾选虚拟化支持，设置概要如图 4.4 所示。

图 4.4 Compute 节点设置

Storage 节点设置概要如图 4.5 所示。

图 4.5　Storage 节点设置

2. VLAN 环境设置

使用模拟器模拟交换机连接 VMware Workstation 的三个网卡，然后进行如下配置，创建 VLAN 来为之后的租户提供独立的网络。

```
sw#vlan database
sw(vlan)#vlan 500 name cloud_public
VLAN 500 added:
    Name: cloud_public
sw(vlan)#vlan 501 name cloud_management
VLAN 501 added:
    Name: cloud_management
sw(vlan)#vlan 502 name cloud_storage
VLAN 502 added:
    Name: cloud_storage
sw(vlan)#vlan 503 name cloud_admin
VLAN 503 added:
    Name: cloud_admin
```

依次添加 VLAN 510～530，支持的设备可以直接输入命令 vlan 510～530 来进行添加。

```
sw#showvlan-switch
VLAN Name                        Status    Ports
---- -------------------------- --------- -------------------------------
1    default                    active    Fa0/0, Fa0/1, Fa0/2, Fa0/3
                                          Fa0/4, Fa0/5, Fa0/6, Fa0/7
                                          Fa0/8, Fa0/9, Fa0/10, Fa0/11
                                          Fa0/12, Fa0/13, Fa0/14, Fa0/15
500  cloud_public               active
```

501	cloud_management	active
502	cloud_storage	active
503	cloud_admin	active
510	VLAN0510	active
512	VLAN0512	active
513	VLAN0513	active
514	VLAN0514	active
515	VLAN0515	active
516	VLAN0516	active
517	VLAN0517	active
518	VLAN0518	active
519	VLAN0519	active
520	VLAN0520	active
521	VLAN0521	active
522	VLAN0522	active
523	VLAN0523	active
524	VLAN0524	active
525	VLAN0525	active
526	VLAN0526	active
527	VLAN0527	active
528	VLAN0528	active
529	VLAN0529	active
530	VLAN0530	active

```
sw#conf t
sw(config)#interface range f0/0 -4
sw(config-if-range)#switchport access vlan 503
sw(config-if-range)#switchport mode access
sw(config-if-range)#spanning-tree portfast
%Portfast has been configured on FastEthernet0/3 but will only
have effect when the interface is in a non-trunking mode.
%Warning: portfast should only be enabled on ports connected to a single host.
 Connecting hubs, concentrators, switches,  bridges, etc.to this interface
whenportfast is enabled, can cause temporary spanning tree loops.
 Use with CAUTION
……
sw(config-if-range)#no shut
sw(config-if-range)#exit
sw(config)#interface range f0/5 -10
sw(config-if-range)#switchport trunk native vlan 500
sw(config-if-range)#switchport trunk allowed vlan 500-503
sw(config-if-range)#switchport mode trunk
*Mar  1 00:44:51.783: %DTP-5-TRUNKPORTON: Port Fa0/5,Fa0/10 has become dot1q trunk
sw(config-if-range)#no shut
sw(config-if-range)#spanning-tree portfasttrunk
%Warning: portfast should only be enabled on ports connected to a single host.
```

```
 Connecting hubs, concentrators, switches, bridges, etc.to this interface
whenportfast is enabled, can cause temporary spanning tree loops.
 Use with CAUTION
……
sw(config-if-range)#exit
sw(config)#interface range f0/11 -15
sw(config-if-range)#switchport trunk allowed vlan 510-530
sw(config-if-range)#switchport mode trunk
sw(config-if-range)#no shut
sw(config-if-range)#spanning-tree portfasttrunk
%Warning: portfast should only be enabled on ports connected to a single host.
 Connecting hubs, concentrators, switches, bridges, etc.to this interface
whenportfast is enabled, can cause temporary spanning tree loops.
 Use with CAUTION
……
sw(config-if-range)#end
sw#wr
Building configuration...
[OK]
```

4.3 部署 OpenStack 环境

部署 OpenStack 使用的镜像可以从 Fuel 官方网站 http://fuel.mirantis.com 注册下载。

1. 安装 Fuel Master 节点

本次安装使用 MirantisOpenStack-6.0 镜像，启动安装时使用系统默认的 IP 地址即可，如图 4.6 所示。如果想更换 IP 地址可以按 Tab 键对默认设置进行修改。

图 4.6　镜像启动界面

整个安装过程无需进行任何操作，过程中会出现很多安装消息，等待时间稍长，请耐心等待。等 Fuel Master 节点安装好后会提示相关登录信息，如图 4.7 所示。

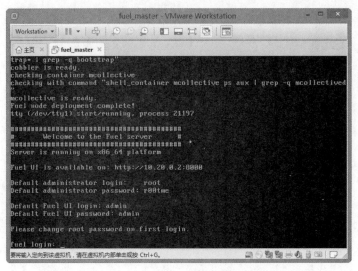

图 4.7　Fuel Master 节点登录信息

因为 Fuel Master 有严格的规则控制，所以根据提示信息直接使用本地浏览器访问地址 http://10.20.0.2:8000 是连接不上 Fuel UI 的，需要通过 Xshell 上的隧道来解决。

首先使用 Xshell 正常连接 Fuel Master 的 22 号端口，然后选择属性中类别下的隧道，建立转移规则，源主机为 localhost，添加目标主机 IP 地址 10.20.0.2 和监听端口号 8000，如图 4.8 所示。

图 4.8　建立隧道

建立隧道后在 Xshell 上使用给定的 root 用户和密码 r00tme 登录字符终端。同时在浏览器地址栏中输入 http://127.0.0.1:8000，使用给定的用户名 admin 以及密码 admin

就可以登录 Fuel Master 的 Web 页面，来进行 OpenStack 环境的配置。Fuel Master 登录页面如图 4.9 所示。

图 4.9　Fuel Master 登录页面

2. 新建 OpenStack 环境

新建 OpenStack 环境需要分别设置 OpenStack 的名称和版本、部署模式、计算、网络、后台存储、附加服务这几项内容。

（1）名称和版本

使用 Fuel Master 的 Web 页面填写新建立的 OpenStack 名称、选择 OpenStack 版本，如图 4.10 所示。这里将安装 CentOS 系统，部署 J 版本的 OpenStack。

图 4.10　选择名称和版本

（2）部署模式

Fuel 部署 OpenStack 环境有两种部署模式，HA 多节点和多节点，使用 HA 多节点部署模式可以避免单点故障，而多节点部署模式也就是非 HA 多节点，不能提供高可用，生产环境中建议使用 HA 多节点部署，这里的虚拟环境中使用多节点即可，如图 4.11 所示。

图 4.11　部署模式

Fuel 支持多种虚拟化管理器类型，比如 KVM、QEMU、vCenter，这里使用的是 VMware 虚拟机环境所以选择 QEMU，如图 4.12 所示。

图 4.12　选择虚拟化管理器类型

Fuel 支持多种网络模式，这里选择使用 Neutron VLAN 模式，如图 4.13 所示。

图 4.13　选择网络模式

最后选择 Fuel 的后端存储，分别选择 Cinder 和 Glance 的存储类型，由于硬件环境有限，所以使用默认的存储类型，如图 4.14 所示。

图 4.14　选择后端存储

附加服务这里就不再选择，如果后期需要部署可以手动添加。这样 OpenStack 环境设置就准备就绪了，继续单击"新建"按钮等待开始部署各节点。

3. 启动待安装虚拟机

启动 Controller、Compute、Storage 节点，等待网卡启动后 Fuel Master 通过 PXE 将 bootstrap 镜像启动，采用默认的第一项（默认为 bootstrap）给三个待部署节点安装基础操作系统，如图 4.15 所示。之后 Fuel Master 节点上的 Cobbler 就给各节点安装基础操作系统，画面出现 bootstrap login 后即部署完毕，三节点的登录用户名和密码都与 Fuel Master 节点一致，为 root/r00tme。

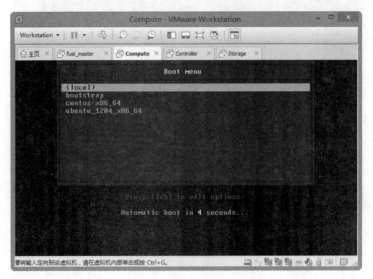

图 4.15　选择安装的系统环境

安装完基础操作系统，Fuel Master 节点会自动发现这些硬件，在 Fuel Web 页面中可以查看被发现的三个节点，如图 4.16 所示。

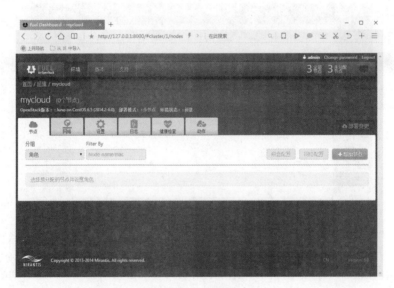

图 4.16　发现的节点

4．配置节点

接下来开始针对发现的三个节点配置 OpenStack 环境，需要分别设置节点角色、配置网络之后部署 OpenStack 的服务。

（1）分配角色

在 Fuel Web 中的节点选项卡单击"增加节点"选项增加一个控制节点，勾选分配角色中的 Controller，在发现的节点中，勾选将设置为控制节点的主机，如图 4.17 所示。

图 4.17　增加控制节点

按照相同的方法分别增加一个计算节点和存储节点，角色分配完毕后如图 4.18 所示。

图 4.18　各节点角色分配情况

之后在节点选项卡，分别勾选各节点，单击"网络配置"选项将节点的物理网卡和 OpenStack 逻辑网络建立映射关系，通过拖拽操作，如图 4.19 所示。

图 4.19　设置节点网卡对应关系

其中 eth0 对应 PXE 网络；eth1 同时承载公共网络、存储网络和管理网络；eth2 对应私有网络。

（2）配置整体网络

配置整体网络是 Fuel 部署中相对复杂的一块，需按照之前规划的网络环境进行部署，在网络选项卡中进行适当的修改，如图 4.20 所示。

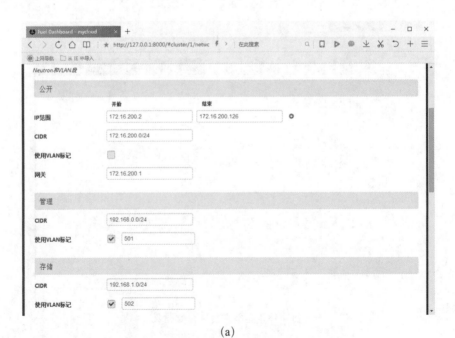

(a)

图 4.20　整体网络

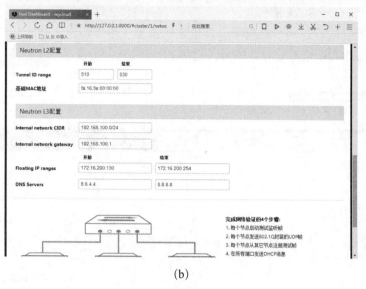

(b)

图 4.20 整体网络（续）

其中位于"公开"部分的网络设置用于物理机与外部网络进行通信。Floating IP 用于动态分配给 OpenStack 的实例，实现和外部网络通信。需要注意的是这里地址段不能重叠。由于公有网络、管理网络和存储网络是共用同一网卡且 IP 地址段不同，要实现二层的隔离，需要打上 vlan 标签，相对应连接的交换机必须启用 trunk 模式，也就是之前在交换机上进行的相关设置。一旦网络配置完毕并安装完成，这个地址是永久不能改变的。

配置完成后单击"验证网络"按钮，检查网络设置是否正确，如果网络配置正确会显示验证成功，如图 4.21 所示，如果网络发生错误，则需要重新检查网络设置情况修正后再次进行验证。

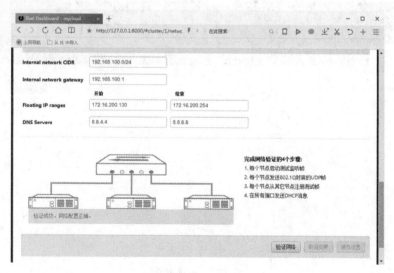

图 4.21 校验网络配置

(3)安装 OpenStack 服务

网络配置完成后返回节点选项卡，单击"部署变更"按钮，开始部署操作系统，三个节点的操作系统全都安装完毕后，会接着安装 OpenStack 服务，如图 4.22 所示。

图 4.22　部署 OpenStack 服务

待所有节点的 OpenStack 服务都安装完毕后，会显示登录到 OpenStack dashboard 页面地址，如图 4.23 所示。

图 4.23　部署完毕

（4）访问 dashboard

访问 dashboard，这里的登录地址是 http://172.16.200.2 或者 http://10.20.0.7。登录界面如图 4.24 所示。

图 4.24　OpenStack 登录界面

使用默认用户名 admin 和密码 admin 进行登录，登录后的主界面如图 4.25 所示。

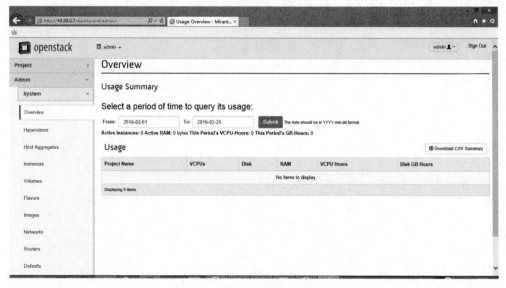

图 4.25　dashboard 主界面

本章总结

- Fuel 是一款快速部署 OpenStack 环境的部署软件，可以根据需要灵活地添加 OpenStack 计算、存储节点。
- Fuel 节点类型分为 Fuel Master、Controller、Compute 和 Storage 几种角色。
- 部署 Fuel 网络，是部署 OpenStack 环境的关键，同时也是 Fuel 部署中相对复杂的部分。

第 5 章

大数据 Hadoop

技能目标

- 了解 Hadoop 体系结构
- 能够安装 Hadoop 运行环境
- 掌握 HDFS 体系结构
- 掌握 HDFS 命令行操作
- 理解 MapReduce 计算模型

本章导读

我们已经进入了大数据（Big Data）时代，数以亿计的用户的互联网服务时时刻刻都在产生巨量的交互，要处理的数据量实在是太大了，以传统的数据库技术手段根本无法满足数据处理的实时性、有效性的要求。

本章将带着大家初步了解 Hadoop 的体系结构，掌握 Hadoop 运行环境与开发环境的安装，初步了解 Hadoop 程序的运行。然后介绍 HDFS 基本原理及常用的 HDFS 管理操作，最后学习 MapReduce 编程框架。

知识服务

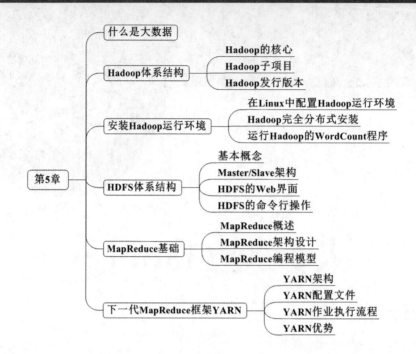

5.1 什么是大数据

大数据是指无法在一定时间内用常规软件工具对其内容进行抓取、管理和处理的数据集合。简而言之就是数据量非常大，大到无法用常规工具处理，如关系型数据库、数据仓库等。这里的"大"又是一个什么量级呢？如阿里巴巴每天所处理的交易数据达到20PB（即20971520GB）。传统数据处理技术为何不能胜任？主要原因是关系型数据库是针对表、字段、行这种可使用二维表格表示的结构化数据而设计的，而大数据通常是针对文本这种非结构化数据。

数据量大是大数据的显著特点，归纳来说大数据特点如下：

- 体量巨大。按目前的发展趋势看，大数据的体量已经达到 PB 甚至 EB 级。
- 大数据的数据类型多样，以非结构化数据为主。如：网络日志、音频、视频、图片、地理位置信息、交易数据、社交数据等。
- 价值密度低。有价值的数据仅占到数据总量相当小的一部分。比如一段监控视频中真正有价值的画面可能只有几秒钟。由于价值密度低，所以迅速地完成数据的价值提纯是目前大数据涌现的背景下亟待解决的难题。
- 产生和要求处理速度快。这是大数据区分于传统数据挖掘最显著的特征。

另外，大数据也是一种方法论。原则是"一切都被记录，一切都被数字化，从数字里寻找需求、寻找知识、发掘价值"，这是一种新的思维方式，不同于此前的专家方式，而是通过数据分析来得到结论，这是大数据时代的一个显著特征。这也就要求技术人员拥有能够从各种各样类型的数据中快速获得有价值信息的能力。

目前有很多大数据处理系统可以处理大数据，如表 5-1 所示。

表 5-1　常见的大数据处理系统

名称	类型	说明
Hadoop	开源	Apache 基金会所开发的分布式系统基础架构。用户可以在不了解分布式底层细节的情况下，开发分布式程序，为本书重点讲解的系统
Spark	开源	类似 Hadoop MapReduce 的并行框架
Stom	开源	实时的、分布式以及具备高容错的计算系统
MongoDB	开源	面向文档的 NoSQL 数据库
IBM PureData	商用	基于 Hadoop，属于 IBM 专家集成系统 PureSystem 家族中的组成部分，主要面向大数据应用
Oracle Exadata	商用	Oracle 的新一代数据库云服务器
SAP Hana	商用	提供高性能的数据查询功能，用户可以直接对大量实时业务数据进行查询和分析
Teradata AsterData	商用	非结构化数据解决方案
EMC GreenPlum	商用	采用了大规模并行处理，支持 50PB 级海量存储与管理
HP Vertica	商用	列式大数据分析数据库

在本门课程中我们学习的是 Hadoop。Hadoop 是开源软件，实现了一个分布式文件系统（Hadoop Distributed File System，HDFS），分布式系统是运行在多个主机上的软件系统。HDFS 有着高容错性的特点，能够自动保存数据的多个副本，并能自动将失败的任务重新分配。Hadoop 设计用来部署在低廉的通用硬件平台上组成集群，提供热插拔的方式加入新的节点来向集群中扩展，将计算任务动态分配到集群中各个节点并保证各节点的动态平衡。总的来说，Hadoop 具有低成本、高扩展性、高效性、高容错性的特点，所以 Hadoop 得到多家厂商支持或采用，包括阿里巴巴、腾讯、百度、Microsoft、Intel、IBM、雅虎等。

5.2　Hadoop 体系结构

Hadoop 源自于 Google 在 2003 到 2004 年公布的关于 GFS（Google File System）、MapReduce 和 BigTable 三篇论文，创始人 Doug Cutting。Hadoop 现在是 Apache 基金会顶级项目，"Hadoop"是一个虚构的名字，是 Doug Cutting 的孩子为其黄色玩具大象所起的名字。

1. Hadoop 的核心

前面提过 HDFS 和 MapReduce 是 Hadoop 的两大核心。通过 HDFS 来实现对分布存储的底层支持，达到高速并行读写与大容量的存储扩展。通过 MapReduce 实现对分布式并行任务处理程序支持，保证高速分析处理数据。HDFS 在 MapReduce 任务处理过程中提供了对文件操作和存储的支持，MapReduce 在 HDFS 的基础上实现了任务的分发、跟踪、执行等工作，并收集结果，二者相互作用，完成了 Hadoop 分布式集群的主要任务。在后面的章节中我们还会详细介绍这两部分。

2. Hadoop 子项目

整个 Hadoop 生态圈已发展成为包含很多子项目的集合。除了两个核心内容外还包括 Hive、Pig、HBase、ZooKeeper 等。比较完整的项目结构如图 5.1 所示。下面分别对它们进行简单介绍。

（1）HDFS：分布式文件系统，整个 Hadoop 体系的基石。

（2）MapReduce/YARN：并行编程模型。YARN 是下一代的 MapReduce 框架，从 Hadoop 0.23.01 版本后，MapReduce 被重构，通常 YARN 也称为 MapReduceV2，老 MapReduce 称为 MapReduce V1。

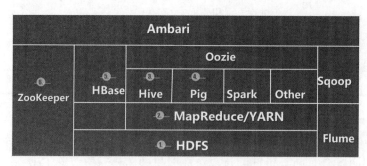

图 5.1　Hadoop 项目结构图

（3）Hive：建立在 Hadoop 上的数据仓库，提供类似 SQL 语言的查询方式查询 Hadoop 中的数据。

（4）Pig：一个对大型数据集进行分析、评估的平台。主要作用类似于数据库里的存储过程。

（5）HBase：全称 Hadoop Database，Hadoop 的分布式的、面向列的数据库，来源于 Google 的关于 BigTable 的论文。主要用于需要随机访问、实时读写的大数据。在后面章节还会详细介绍。

（6）ZooKeeper：是一个为分布式应用所设计的协调服务。主要为用户提供同步、配置管理、分组和命名等服务，减轻分布式应用程序所承担的协调任务。

当然还有大量其他项目加入到 Hadoop 生态圈，如：

- Sqoop：主要用于 Hadoop 与传统数据库（MySQL 等）间的数据传递。
- Flume：日志采集系统。
- Spark：前面提过，是一个相对独立于 Hadoop 的大数据处理系统，可单独进行分布式处理，在这里列出来是因为它可以和 HDFS 很好地结合。
- Oozie：可以将多个 MapReduce 作业组合到一个逻辑工作单元中，进行工作计划的安排，类似于工作流管理引擎。

- Ambari：支持 Hadoop 集群的管理、监控的 Web 工具。

经过近十多年的发展，越来越多的项目加入到了 Hadoop 的生态圈，在本课程中着重介绍 3 个模块，分别是：HDFS、MapReduce、HBase。对于 Hadoop 应用开发人员来说，这也是最基本的模块。

3．Hadoop 发行版本

Hadoop 的版本比较混乱，在这里有必要帮助大家理清一下思路，也有助于大家在参考其他资料时注意区别版本间的差异。总的来说，Hadoop 分为两代，如表 5-2 所示。

表 5-2 Hadoop 版本说明

Apache Hadoop	大版本	说明
第二代 Hadoop 2.0	2.x.x	下一代 Hadoop 由 0.23.x 演化而来
	0.23.x	下一代 Hadoop
第一代 Hadoop 1.0	1.0.x	稳定版，由 0.20.x 演化而来
	0.22.x	非稳定版
	0.21.x	非稳定版
	0.20.x	经典版本，最后演化成 1.0.x

由表可以看出，通常我们所说的 Hadoop 1.0 指的是 0.20.x、0.21.x、0.22.x、1.0.x，Hadoop 2.0 指的是 0.23.x 、2.x.x。在本书中，我们将选择 Hadoop 2.6.0 进行示例讲解。

第二代 Hadoop 一个重大的变化就是重构了 MapReduce，在后面我们就会看到新老 API 同时存在的情况。Hadoop 的下载地址：http://hadoop.apache.org/releases.html。

5.3 安装 Hadoop 运行环境

5.3.1 在 Linux 中配置 Hadoop 运行环境

这里以 CentOS 6.5（64 位）为例进行讲解。

1．用户创建

为方便后续示例统一演示，先在 Linux 中新建用户 hduser。新建用户需要切换至 root 用户登录，root 密码与虚拟机初始安装用户密码相同。

（1）创建 hadoop 用户组，输入命令：

```
groupadd hadoop
```

（2）创建 hduser 用户，输入命令：

```
useradd -g hadoop hduser
```

（3）设置 hduser 的密码，输入命令：

```
passwd hduser
```

按提示输入 2 次密码。

（4）为 hduser 用户添加权限，输入命令：

```
# 修改权限
chmod 777 /etc/sudoers
# 编辑 sudoers
gedit /etc/sudoers
# 还原默认权限
chmod 440 /etc/sudoers
```

先修改 sudoers 文件权限，并在文本编辑窗口中查找行"root ALL=(ALL) ALL"，紧跟后面新加行"hduser ALL=(ALL) ALL"，将 hduser 添加到 sudoers。添加完成后切记还原默认权限，否则系统将不允许使用 sudo 命令。

（5）设置好后重启虚拟机：

```
sudo reboot
```

重启后切换到 hduser 用户登录。

2. 安装 JDK

（1）下载 jdk-7u67-linux-x64.rpm，并进入下载目录。

（2）运行安装命令：

```
sudo rpm -ivh jdk-7u67-linux-x64.rpm
```

完成后查看安装路径，输入命令：

```
rpm -qa jdk -l
```

记住该路径，如"/usr/java/jdk1.7.0.67"。

（3）配置环境变量，输入命令：

```
sudo gedit/etc/profile
```

打开 profile 文件，在文件最下面加入如下内容：

```
export JAVA_HOME=/usr/java/jdk1.7.0.67
export CLASSPATH=$ JAVA_HOME/lib:$CLASSPATH
export PATH=$ JAVA_HOME/bin:$PATH
```

保存后关闭文件，然后输入命令使环境变量生效：

```
source /etc/profile
```

（4）验证 JDK，输入命令：

```
java -version
```

若出现正确版本号，表示安装成功。

3. 配置本机 SSH 免密码登录

（1）使用 ssh-keygen 生成私钥与公钥文件：

```
ssh-keygen -t rsa
```

将在 /home/hduser/.ssh/ 目录下生成 id_rsa（私钥）和 id_rsa.pub（公钥）两个文件。

（2）私钥留在本机，公钥分发给其他主机（现在是 localhost）。输入命令：

```
ssh-copy-id localhost
```

首次运行会要求输入被连接主机的密码，之后就可以免密码登录。命令完成后将在被连接主机的 /home/hduser/.ssh/ 下生成 authorized_key 文件，其中记录了收到的所有来自其他主机的公钥。

（3）现在就可免密码登录了。客户端（发起连接请求）利用私钥签名，服务端（接受连接请求）使用公钥来认证。输入命令：ssh localhost。

其中 localhost 为主机 hostname 或者 IP 地址。

4. 配置其他主机 SSH 免密码登录

与配置本机 SSH 免密码登录过程相同，只不过将 localhost 换成目标主机的 hostname（主机名）或 IP。

为了实现与其他主机 SSH 免密码登录，将已经安装配置完成的虚拟机标识为 node1，将 node1 克隆 2 次，然后配置到克隆出来的主机的免密码登录，最后形成 3 台完全一样的虚拟机，并且都已经安装了 JDK 及 OpenSSH，用户均为 hduser。3 台主机分别标识为 node1、node2（克隆机）、node3（克隆机）。步骤如下：

（1）克隆 2 次。

（2）分别启动并进入 3 台虚拟机，使用 ifconfig 查看主机各自 IP 地址。

（3）修改每台主机的 hostname 及 hosts 文件。

（4）我们已经在 node1 上生成过密钥对，所以现在只要在 node1 上输入命令：

```
ssh-copy-id node2
ssh-copy-id node3
```

这样便可将 node1 的公钥发布到 node2、node3。

（5）测试 SSH，在 node1 上输入命令：

```
ssh node2
# 退出登录
exit
ssh node3
exit
```

到此，我们可以从 node1 免密码登录到 node2 和 node3，这对于 Hadoop 来说就足够了，因为后面我们将 node1 作为 Master 节点（主节点，管理者角色），其余主机为 Slave 节点（从节点），所以不用再配置从 node2、node3 免密码登录到 node1。

5.3.2 Hadoop 完全分布式安装

Hadoop 有三种运行方式:
- 单机模式:无须配置,Hadoop 被视为一个非分布式模式运行的独立 Java 进程。
- 伪分布式:只有一个节点的集群,这个节点既是 Master(主节点、主服务器)也是 Slave(从节点、从服务器),可在此单节点上以不同的 Java 进程模拟分布式中的各类节点。
- 完全分布式:对于 Hadoop,不同的系统会有不同的节点划分方式。在 HDFS 看来分为 NameNode(管理者)和 DataNode(工作者),其中 NameNode 只有一个,DataNode 可有多个;在 MapReduce 看来节点又分为 JobTracker(作业调度者)和 TaskTracker(任务执行者),其中 JobTracker 只有一个,TaskTracker 可以有多个。NameNode 和 JobTracker 可以部署在不同的机器上,也可以部署在同一机器上。部署 NameNode 和 JobTracker 的机器是 Master,其余的机器都是 Slave。

单机模式和伪分布式均不能体现云计算的优势,通常用于程序测试与调试。所以这里以前面已经配置完成的 3 台主机介绍 Hadoop 完全分布式模式的安装。主机名、IP 及角色的对应关系如表 5-3 所示。大家在安装时,注意区分自己的主机名与相应 IP 地址。

表 5-3 主机名与对应的 IP 地址、角色

主机名	IP 地址	所分配的角色
node1	192.168.70.130	Master,NameNode,JobTracker
node2	192.168.70.131	Slave,DataNode,TaskTracker
node3	192.168.70.132	Slave,DataNode,TaskTracker

具体的安装过程如下。

1. 安装 Hadoop

(1)获取 Hadoop 压缩包 hadoop-2.6.0.tar.gz,下载后可使用 VMware Tools 通过共享文件夹或者使用 Xftp 工具传送到 node1。进入 node1,将压缩包解压到 /home/hduser 目录下:

```
# 进入 HOME 目录即:"/home/hduser"
cd ~
tar -zxvf hadoop-2.6.0.tar.gz
```

(2)重命名为 hadoop:

```
mv hadoop-2.6.0 hadoop
```

进入"hadoop"目录,其目录结构如下:
- bin:执行文件目录。包含 hadoop、dfs、yarn 等命令,所有用户均可执行。
- etc:Hadoop 配置文件都在此目录。

- include：包含 C 语言接口开发所需头文件。
- lib：包含 C 语言接口开发所需链接库文件。
- libexec：运行 sbin 目录中的脚本会调用该目录下的脚本。
- logs：日志目录，在运行过 Hadoop 后会生成该目录。
- sbin：仅超级用户能够执行的脚本，包括启动和停止 dfs、yarn 等。
- share：包括 doc 和 hadoop 两个目录，其中 doc 目录包含大量的 Hadoop 帮助文档；hadoop 目录包含了运行 Hadoop 所需的所有 jar 文件，在开发中用到的 jar 文件也可在该目录找到。

（3）配置 hadoop 环境变量，将以下 3 行脚本增加到 profile 内：

```
#hadoop
export HADOOP_HOME=/home/hduser/hadoop
export PATH=$HADOOP_HOME/bin:$PATH
```

保存关闭，最后输入命令使配置生效：

```
source /etc/profile
```

 注意

node2、node3 都要按以上步骤配置。

2. 配置 Hadoop

先进入 node1 进行配置，随后将配置文件复制到 node2、node3。配置 Hadoop 主要涉及的配置文件有 7 个，都在 hadoop/etc/hadoop 文件夹下，可以使用 gedit 命令对其进行编辑。包括：hadoop-env.sh、yarn-env.sh、slaves、core-site.xml、hdfs-site.xml、mapred-site.xml、yarn-site.xml。具体配置如下：

（1）hadoop-env.sh 文件用于指定 JDK 路径。首先打开该文件：

```
[hduser@node1 ~]$ cd ~/hadoop/etc/hadoop
[hduser@node1 hadoop]$ gedit hadoop-env.sh
```

然后增加如下内容指定 JDK 路径：

```
export JAVA_HOME=/usr/java/jdk1.7.0_67
```

（2）yarn-env.sh：第二代 Hadoop 新增加的 YARN 框架（MapReduce V2），针对 MapReduceV1 存在的缺陷进行了重构，使用户基于 MapReduce V1 编写的程序无需修改也可以运行在 YARN 中。将文件打开后指定 JDK 路径：

```
export JAVA_HOME=/usr/java/jdk1.7.0_67
```

（3）slaves：用于增加 Slave 节点即 DataNode 节点。打开并清空原内容，然后输入如下内容：

```
node2
node3
```

表示 node2、node3 作为 Slave 节点。

（4）core-site.xml：该文件是 Hadoop 全局配置。打开并在 <configuration> 元素中增加配置属性如下：

```xml
<configuration>
    <property>
        <name>fs.defaultFS</name>
        <value>hdfs://node1:9000</value>
    </property>
    <property>
        <name>hadoop.tmp.dir</name>
        <value>file:/home/hduser/hadoop/tmp</value>
    </property>
</configuration>
```

这里给出了两个常用的配置属性。fs.defaultFS 表示客户端连接 HDFS 时，默认的路径前缀，9000 是 HDFS 工作的端口。hadoop.tmp.dir 如不指定会保存到系统的默认临时文件目录 /tmp 中。

（5）hdfs-site.xml：该文件是 HDFS 的配置。打开并在 <configuration> 元素中增加配置属性如下：

```xml
<configuration>
    <property>
        <name>dfs.namenode.secondary.http-address</name>
        <value>node1:50090</value>
    </property>
    <property>
        <name>dfs.namenode.name.dir</name>
        <value>file:/home/hduser/hadoop/dfs/name</value>
    </property>
    <property>
        <name>dfs.datanode.data.dir</name>
        <value>file:/home/hduser/hadoop/dfs/data</value>
    </property>
    <property>
        <name>dfs.replication</name>
        <value>2</value>
    </property>
    <property>
        <name>dfs.webhdfs.enabled</name>
        <value>true</value>
    </property>
</configuration>
```

参数具体描述如表 5-4 所示。

表 5-4　hdfs-site.xml 的配置

参数名（name）	描述
dfs.namenode.secondary.http-address	Secondary NameNode 服务器 HTTP 地址和端口
dfs.namenode.name.dir	NameNode 存储名字空间及汇报日志的位置
dfs.datanode.data.dir	DataNode 存放数据块的目录列表
dfs.replication	冗余备份数量，一份数据可设置多个备份
dfs.webhdfs.enabled	在 NameNode 和 DataNode 中启用 WebHDFS

（6）mapred-site.xml：该文件是 MapReduce 的配置，可从模板文件 mapred-site.xml.template 复制。打开并在 <configuration> 元素中增加配置属性如下：

```
<configuration>
    <property>
        <name>mapreduce.framework.name</name>
        <value>yarn</value>
    </property>
    <property>
        <name>mapreduce.jobhistory.address</name>
        <value>node1:10020</value>
    </property>
    <property>
        <name>mapreduce.jobhistory.webapp.address</name>
        <value>node1:19888</value>
    </property>
</configuration>
```

重点是这里使用 mapreduce.framework.name 配置属性指定了使用 YARN 框架运行 MapReduce 程序。

（7）yarn-site.xml：如果在 mapred-site.xml 配置了使用 YARN 框架，那么 YARN 框架使用此文件中的配置。打开并在 <configuration> 元素中增加配置属性如下：

```
<configuration>
    <property>
        <name>yarn.nodemanager.aux-services</name>
        <value>mapreduce_shuffle</value>
    </property>
    <property>
        <name>yarn.nodemanager.aux-services.mapreduce.shuffle.class</name>
        <value>org.apache.hadoop.mapred.ShuffleHandler</value>
    </property>
    <property>
        <name>yarn.resourcemanager.address</name>
```

```xml
            <value>node1:8032</value>
    </property>
    <property>
            <name>yarn.resourcemanager.scheduler.address</name>
            <value>node1:8030</value>
    </property>
    <property>
            <name>yarn.resourcemanager.resource-tracker.address</name>
            <value>node1:8035</value>
    </property>
    <property>
            <name>yarn.resourcemanager.admin.address</name>
            <value>node1:8033</value>
    </property>
    <property>
            <name>yarn.resourcemanager.webapp.address</name>
            <value>node1:8088</value>
    </property>
</configuration>
```

YARN 的配置属性在后面章节有详细讲解，此处先大概了解。

最后将这 7 个文件复制到 node2、node3 的同目录下。最简单的方式是使用基于 SSH 的 scp 命令，在 node1 上输入如下命令：

```
scp -r /home/hduser/hadoop/etc/hadoop/ hduser@node2:/home/hduser/hadoop/etc/
scp -r /home/hduser/hadoop/etc/hadoop/ hduser@node3:/home/hduser/hadoop/etc/
```

以上两条命令分别表示将 node1 的 Hadoop 配置文件夹复制到 node2、node3。

3. 验证

下面验证 Hadoop 配置是否正确。

（1）在 Master 主机（node1）上格式化 NameNode：

```
[hduser@node1 ~]$ cd ~/hadoop
[hduser@node1 hadoop]$ bin/hdfs namenode -format
```

（2）关闭 node1、node2、node3 系统防火墙并重启虚拟机：

```
service iptables stop
sudochkconfig iptables off
reboot
```

（3）启动 HDFS：如图 5.2 所示。

```
[hduser@node1 ~]$ cd ~/hadoop
[hduser@node1 hadoop]$ sbin/start-dfs.sh
```

```
[hduser@node1 hadoop]$ ./sbin/start-dfs.sh
16/05/23 16:18:36 WARN util.NativeCodeLoader: Unable to load native-hadoop library for yo
ur platform... using builtin-java classes where applicable
Starting namenodes on [node1]
node1: starting namenode, logging to /home/hduser/hadoop/logs/hadoop-hduser-namenode-node
1.out
node3: starting datanode, logging to /home/hduser/hadoop/logs/hadoop-hduser-datanode-node
3.out
node2: starting datanode, logging to /home/hduser/hadoop/logs/hadoop-hduser-datanode-node
2.out
Starting secondary namenodes [node1]
node1: starting secondarynamenode, logging to /home/hduser/hadoop/logs/hadoop-hduser-seco
ndarynamenode-node1.out
16/05/23 16:18:57 WARN util.NativeCodeLoader: Unable to load native-hadoop library for yo
ur platform... using builtin-java classes where applicable
[hduser@node1 hadoop]$
```

图 5.2 启动 HDFS

（4）输入 jps 命令查看 Java 进程，如图 5.3 所示。

```
node1: starting namenode, logging to /home/hduser/hadoop/logs/hadoop-hduser-namenode-node
1.out
node3: starting datanode, logging to /home/hduser/hadoop/logs/hadoop-hduser-datanode-node
3.out
node2: starting datanode, logging to /home/hduser/hadoop/logs/hadoop-hduser-datanode-node
2.out
Starting secondary namenodes [node1]
node1: starting secondarynamenode, logging to /home/hduser/hadoop/logs/hadoop-hduser-seco
ndarynamenode-node1.out
16/05/23 16:18:57 WARN util.NativeCodeLoader: Unable to load native-hadoop library for yo
ur platform... using builtin-java classes where applicable
[hduser@node1 hadoop]$ jps
5901 Jps
5607 NameNode
5792 SecondaryNameNode
[hduser@node1 hadoop]$
```

图 5.3 查看 Java 进程

（5）启动 YARN：

[hduser@node1 hadoop]$ sbin/start-yarn.sh

也可以使用 start-all.sh 同时启动 HDFS 和 YARN。

（6）查看集群状态：

[hduser@node1 hadoop]$ bin/hdfs dfsadmin -report

运行该命令后，输出如下信息则表示 Hadoop 已成功运行。

[hduser@node1 hadoop]$ bin/hdfs dfsadmin -report
16/05/23 16:22:35 WARN util.NativeCodeLoader: Unable to load native-hadoop library for your
　　platform... using builtin-java classes where applicable
Configured Capacity: 37558796288 (34.98 GB)
Present Capacity: 28704219136 (26.73 GB)
DFS Remaining: 28704161792 (26.73 GB)
DFS Used: 57344 (56 KB)
DFS Used%: 0.00%
Under replicated blocks: 0
Blocks with corrupt replicas: 0
Missing blocks: 0

Live datanodes (2):

Name: 192.168.70.131:50010 (node2)
Hostname: node2
Decommission Status : Normal
Configured Capacity: 18779398144 (17.49 GB)
DFS Used: 28672 (28 KB)
Non DFS Used: 4421865472 (4.12 GB)
DFS Remaining: 14357504000 (13.37 GB)
DFS Used%: 0.00%
DFS Remaining%: 76.45%
Configured Cache Capacity: 0 (0 B)
Cache Used: 0 (0 B)
Cache Remaining: 0 (0 B)
Cache Used%: 100.00%
Cache Remaining%: 0.00%
Xceivers: 1
Last contact: Mon May 23 16:22:34 CST 2016

Name: 192.168.70.132:50010 (node3)
Hostname: node3
Decommission Status : Normal
Configured Capacity: 18779398144 (17.49 GB)
DFS Used: 28672 (28 KB)
Non DFS Used: 4432711680 (4.13 GB)
DFS Remaining: 14346657792 (13.36 GB)
DFS Used%: 0.00%
DFS Remaining%: 76.40%
Configured Cache Capacity: 0 (0 B)
Cache Used: 0 (0 B)
Cache Remaining: 0 (0 B)
Cache Used%: 100.00%
Cache Remaining%: 0.00%
Xceivers: 1
Last contact: Mon May 23 16:22:36 CST 2016

（7）在浏览器中查看 HDFS 运行状态：

http://node1:50070

（8）停止 Hadoop：

[hduser@node1 hadoop]$ sbin/stop-all.sh

5.3.3 运行 Hadoop 的 WordCount 程序

WordCount 程序是 Hadoop 本身提供的使用 MapReduce 框架编写的入门程序，类似于刚学习 Java 语言时编写的 Hello World 程序。WordCount 实现了对文本中的单词计数的功能，并要求输出结果并按单词首字母排序。比如，输入一个文件，其内容如下：

hello world
hello hadoop
hello mapreduce

输出结果为：

hadoop　　　　1
hello　　　　　3
mapreduce　　1
world　　　　 1

本小节暂不介绍如何编写 MapReduce，仅是运行该 MapReduce 程序来测试 Hadoop 的运行情况。首先我们要找出 WordCount 程序，然后我们还需要提供被计数的文本文件。Hadoop 的示例程序位于 share/hadoop/mapreduce/hadoop-mapreduce-examples-2.6.0.jar 中，其中就包含了 WordCount 程序。接下来创建 2 个文本文件并保存到 Hadoop HDFS 中，本小节以下部分涉及到的命令在后续章节还会详细介绍，在此可先简单了解 Hadoop 相关的 shell 操作。

（1）在操作系统 /home/hduser/file 目录下创建 file1.txt、file2.txt，可使用图形界面创建。

file1.txt 输入内容：Hello World hi HADOOP
file2.txt 输入内容：Hello hadoop hi CHINA

（2）启动 HDFS 后创建 HDFS 目录 /input2：

[hduser@node1 hadoop]$bin/hadoop fs -mkdir /input2

（3）将 file1.txt、file2.txt 保存到 HDFS 中：

[hduser@node1 hadoop]$bin/hadoop fs -put ~/file/file*.txt/input2/

（4）查看 HDFS 上是否已经存在 file1.txt、file2.txt：

[hduser@node1 hadoop]$bin/hadoop fs -ls /input2

如图 5.4 所示表示 HDFS 保存文件成功。

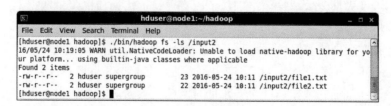

图 5.4　查看 HDFS 目录

有了示例程序及文本文件，就可以通过运行包命令"hadoop jar xxx.jar"来执行 WordCount 程序。进入 node1 主机上的 Hadoop 安装目录，输入如下命令：

[hduser@node1 hadoop]$bin/hadoop jar share/hadoop/mapreduce/hadoop-mapreduce-examples-2.6.0.jar wordcount /input2/ /output2/wordcount1

"wordcount"为示例 jar 包提供的参数选项，wordcount 仅是其提供功能之一；"/input2/"表示文本文件输入目录（HDFS）；"/output2/wordcount1"表示提供的输出目录（HDFS）参数。可以发现，在数据量小的时候，使用 Hadoop 运行并没有优势，花费在通信同步上的开销比实际单词统计处理开销还要多，所以说 Hadoop 主要用于大数据的处理。当执行完成后，使用如下命令来查看输出目录中所有结果：

[hduser@node1 hadoop]$bin/hadoop fs -cat /output2/wordcount1/*

结果如图 5.5 所示。表示已经成功安装了 Hadoop。

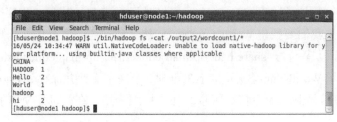

图 5.5　查看单词统计结果

5.4　HDFS 体系结构

前面提过，大数据的特点是体量大、类型繁多、价值密度低、产生和处理速度快。当数据集的大小超过一台独立的物理计算机的存储能力时，就必须对它进行分区并存储到若干台单独的计算机上，将这些计算机通过网络连接，并对网络中的文件系统进行集中管理，由此构成分布式文件系统。HDFS 便是这样的分布式文件系统，可以将 HDFS 的优势归纳如下：

- 存储超大文件，文件大小通常都是上百 MB、TB、PB 级别。
- 标准流式访问，基于"一次写入，多次读取"的构建思路。即只支持文件的追加写，不支持随机访问，这是最高效的访问模式。往往每次对数据的分析都涉及大部分数据甚至全部，所以在设计上优先加快整个数据集的访问，而非单条数据记录。
- 运行在廉价的商用机器集群上，Hadoop 并不需要昂贵且高可靠的硬件。

同样，HDFS 也存在缺点，以下应用就不适合在 HDFS 上运行：

- 低时间延迟的数据访问。要求低时间延迟数据访问的应用，例如几十毫秒范围，不适合在 HDFS 上运行。HDFS 虽有着高数据吞吐量，但是以提高时间延迟为代价。后续将会介绍的 HBase 可以满足低延迟的访问需求。
- 无法高效存储大量小文件，大量的小文件会造成整个文件系统的目录树及索引目录相对变大，而这些都是存放在 NameNode 节点。

- 不支持多用户写入及任意修改文件。由于是基于流式访问，目前在 HDFS 的一个文件中只能有一个用户写入，并且写操作只能在文件末尾完成。但是随着后续版本的不断更新，多用户对同一文件写操作在以后可能会实现。

HDFS 的体系结构如图 5.6 所示，下面针对此图中各部分分别介绍。

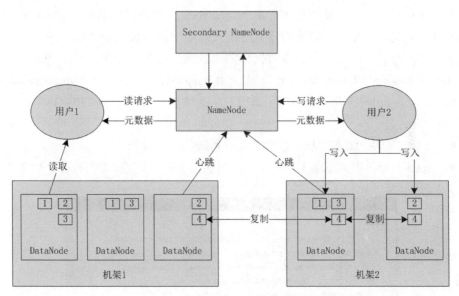

图 5.6　HDFS 体系结构

5.4.1　基本概念

1. 文件块

HDFS 中的文件被分成块进行存储，如图中 DataNode 中的以数字编号的方块。它是文件存储处理的最小逻辑单元，默认块大小为 64MB，使用文件块的好处是：

- 文件的所有块并不需要存储在同一个磁盘上，可以利用集群上的任意一个磁盘进行存储。
- 对分布式系统来说，由于块的大小是固定的，因此计算单个磁盘能存储多少个块就相对容易，可简化存储管理。
- 在数据冗余备份时，将每个块复制到少数几台独立的机器上（默认为 3 台），可以确保在块、磁盘或机器发生故障后数据不会丢失。如果发现一个块不可用，系统会从其他地方读取另一个副本，这个过程对用户是透明的。

使用 fsck 命令可以显示块信息：

语法：

hdfs fsck / -files -blocks

fsck 命令将列出文件系统中根目录"/"下各个文件由哪些块构成。fsck 命令只是从 NameNode 获取信息，并不与任何 DataNode 交互，因此并不真正获取数据。

2. NameNode 和 DataNode 节点

NameNode 和 DataNode 属于 HDFS 集群的两类节点。

NameNode 负责管理文件系统的命名空间，属于管理者角色。它维护文件系统树内所有文件和目录，记录每个文件在各个 DataNode 上的位置和副本信息，并协调客户端对文件的访问。这些信息以两种形式存在：命名空间镜像文件（fsimage_*）和编辑日志文件（edits_*），用于存储跟数据相关的数据，称为元数据，位于 NameNode 节点 node1 的 hadoop/dfs/name/current 目录（由 hdfs-site.xml 中的 dfs.namenode.name.dir 属性指定）下，如图 5.7 所示。由图可看出除了元数据，还包括"VERSION"和"seen_txid"两个文件。

- VERSION：版本信息，包含了文件系统的唯一标识符。
- seen_txid：该文件用于事务管理，里面保存一个整数，表示 edits_* 的尾数。

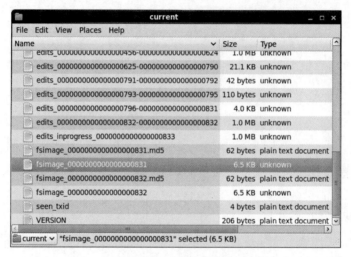

图 5.7 Hadoop NameNode 节点中的元数据

DataNode 根据需要存储并检索数据块，并定期向 NameNode 发送所存储的块的列表，属于工作者角色。负责所在物理节点的的存储管理，按照一次写入，多次读取的原则。存储文件由数据块组成，典型的块大小是 64MB，尽量将各数据块分布到各个不同的 DataNode 节点上。在图 5.1 中，某个文件被分成 4 块，在多个 DataNode 中存储，并且每块复制 2 个副本，存储在其他 DataNode 上。DataNode 节点的数据存储目录为 /home/hduser/hadoop/dfs/data（由 hdfs-site.xml 中的 dfs.datanode.data.dir 属性指定）。以 node2 中部分数据块的某一目录为例，如 "/home/hduser/hadoop/dfs/data/current/BP-367303913-192.168.70.130-1463549699942/current/finalized/subdir0/subdir0"，该目录下文件内容如图 5.8 所示。主要包括两类文件：

- blk_<id>：HDFS 的数据块，保存具体的二进制数据。
- blk_<id>.meta：数据块的属性信息，包括版本信息、类型信息等。

图 5.8　DataNode 文件解析

DataNode 负责处理文件系统客户端的文件读写请求，并在 NameNode 的统一调度下进行数据块的创建、删除和复制工作。如果在 NameNode 上的数据损坏，HDFS 中所有的文件都不能被访问，由此可见 NameNode 节点的重要性。为了保证 NameNode 的高可用性，Hadoop 对 NameNode 进行了补充，即 Secondary NameNode 节点。

3. Secondary NameNode 节点

系统中会同步运行一个 Secondary NameNode，也称二级 NameNode。相当于 NameNode 的快照，能够周期性地备份 NameNode，记录 NameNode 中的元数据等，可以用来恢复 NameNode，但 Secondary NameNode 中的备份会滞后于 NameNode，所以会带来一定的数据损失。为了防止宕机，通常是将 Secondary NameNode 和 NameNode 设置为不同的主机。使用 hdfs-site.xml 中配置的 dfs.namenode.secondary.http-address 属性值可以通过浏览器查看 Secondary NameNode 的运行状态，如图 5.9 所示。

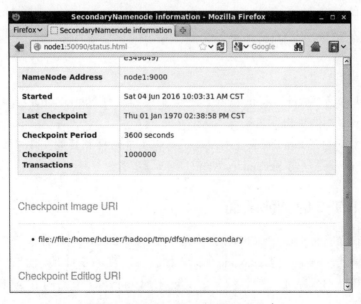

图 5.9　Secondary NameNode 运行状态

5.4.2 Master/Slave 架构

一个 HDFS 集群是由一个 NameNode 和多个 DataNode 组成，属于典型的 Master/Slave 模式。如图 5.1 所示，其读写流程如下：

- 数据读流程：由客户端向 NameNode 请求访问某个文件，NameNode 返回该文件所在位置即在哪个 DataNode 上，然后由客户端从该 DataNode 读取数据。
- 数据写流程：由客户端向 NameNode 发出文件写请求，NameNode 告诉客户该向哪个 DataNode 写入文件，然后由客户将文件写入 DataNode 节点，随后 DataNode 将自动复制到其他 DataNode 节点上，默认 3 份备份。

了解了 HDFS 的节点之间的结构关系后，下面介绍 HDFS 常用的节点管理操作。

1. 节点添加

可扩展性是 HDFS 的一个重要特性，向 HDFS 集群中添加节点很容易实现。添加一个新的 DataNode 节点步骤如下。

（1）首先对新节点进行系统配置，包括 hostname、hosts 文件、JDK 环境、防火墙等。

（2）然后在新加节点上安装好 Hadoop，要和 NameNode 使用相同的配置，可以直接从 NameNode 复制。

（3）然后在 NameNode 上修改 $HADOOP_HOME/conf/slaves 文件，加入新加节点主机名。

（4）运行启动命令：bin/start-all.sh。

2. 负载均衡

HDFS 的数据在各个 DataNode 中的分布可能很不均匀，尤其是在 DataNode 节点出现故障或新增 DataNode 节点时。使用如下命令可重新平衡 DataNode 上的数据块的分布：

```
sbin/start-balancer.sh
```

3. 安全机制

Master/Slave 架构通过 NameNode 来统一调度，没有 NameNode，文件系统将无法使用。Hadoop 采用两种机制来确保 NameNode 的安全。

- 第一种是将 NameNode 上存储的元数据文件转移到其他文件系统中。
- 第二种就是使用 Secondary NameNode 同步备份。

5.4.3 HDFS 的 Web 界面

通过 http://NameNodeIP:50070 可以访问 HDFS 的 Web 界面，该界面提供了 NameNode 基本信息与所有 DataNode 信息，如图 5.10 和图 5.11 所示。在后续章节中，"NameNodeIP"不作特别说明情况下，其表示 NameNode 节点的 IP 或 hostname，请自行替换。

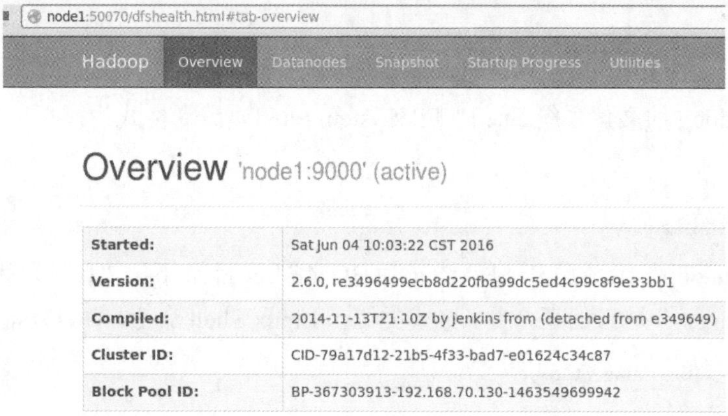

图 5.10　NameNode 基本信息

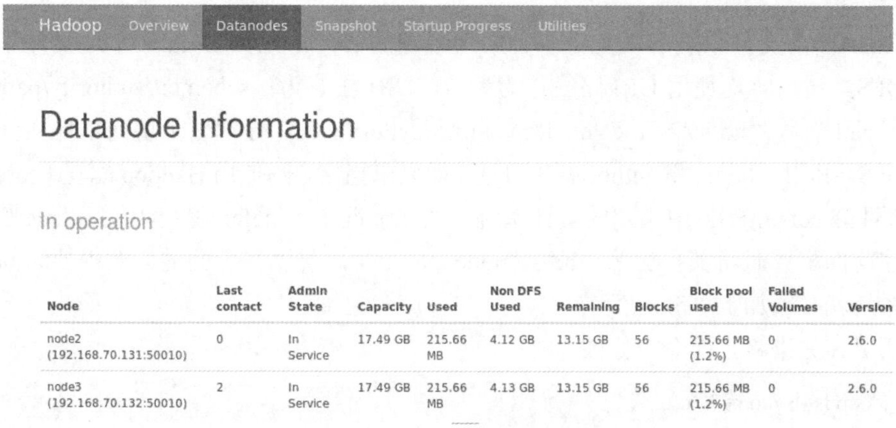

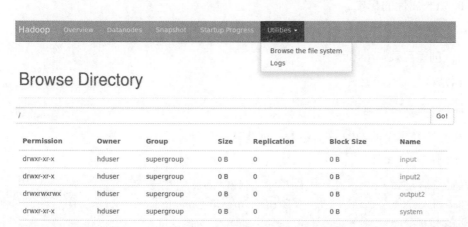

图 5.11　DataNode 基本信息

此外还提供浏览文件系统、日志功能等，如图 5.12 所示。

图 5.12　浏览文件系统

5.4.4 HDFS 的命令行操作

调用 Hadoop 的文件系统 Shell（FileSystem Shell）命令格式为：
语法：

```
hadoop fs <args>
```

其中"hadoop"命令位于 $HADOOP_HOME/bin 目录下，"fs"为其参数，表示 FS Shell，"<args>"是 fs 的子命令，格式类似于 Linux Shell 命令，并且功能也类似。如：

```
hadoop fs -ls file:///home/hduser
```

该命令使用了 fs 子命令 -ls，最后一个参数表示本地系统文件路径。运行命令后会列出该目录下所有文件。可以看出 Hadoop fs 子命令 -ls 与 Linux 中的 ls 命令的作用及语法均是相似的。但应该注意到，在 HDFS 中访问本地系统文件加上了前缀"file://"，那么是否可以省略呢？

Hadoop 的文件系统可以支持多种文件系统的访问，比如 Local（本地文件系统）和 HDFS。访问时均使用 URI 路径作为参数，URI 格式为：scheme://authority/path，如"file://path"和"hdfs://NameNodeIP:NameNodePort/path"分别表示访问本地文件系统和 HDFS。其中 scheme 和 authority 为可选，如果未指定，则使用 Hadoop 配置（core-site.xml 属性 fs.defaultFS）中指定的文件系统，默认情况下 fs.defaultFS 的值为"file:///"，由于在前面配置时将其指定为"hdfs://node1:9000"，所以在上面的命令中要访问本地文件系统就必须加上"file://"前缀。

所以，如果将上面命令修改为：

```
hadoop fs -ls /home/hduser
```

则表示查看 HDFS 文件系统下的"/home/hduser"目录，如果这个目录在 HDFS 中不存在就会失败。在后续讲解中，不作特别说明我们会省略 URI 前缀，请注意在环境中 fs.defaultFS 属性的配置。

了解基本的语法后，再具体来看 fs 操作常用子命令。

1. 创建目录：mkdir

语法：

```
hadoop fs -mkdir <paths>
```

接受路径指定的 URI 作为参数，创建这些目录。
示例：

```
hadoop fs -mkdir /user              # 在 HDFS 中创建"/user"目录
hadoop fs -mkdir /user/hadoop       # 在 HDFS 中创建"/user/hadoop"目录
hadoop fs -mkdir /user/hadoop/dir1 /user/hadoop/dir2# 同时创建多个目录
```

2. 列表文件：ls

语法：

hadoop fs -ls <args>

如果是文件，则按照如下格式返回文件信息：

文件名 <副本数> 文件大小 修改日期 修改时间 权限用户 ID 组 ID

如果是目录，则返回它直接子文件的一个列表。

3. 查看文件：cat

语法：

hadoop fs -cat URI [URI…]

输出路径指定文件的内容。

示例：

hadoop fs -cat /input2/file1.txt /input2/file2.txt # 查看 HDFS 文件 file1.txt 和 file2.txt
hadoop fs -cat file:///file3 # 查看本地系统文件 /file3

4. 转移文件：put、get、mv、cp

（1）put 命令

语法：

hadoop fs -put <localsrc>...<dst>

从本地文件系统中复制单个或多个文件到 HDFS，也支持从标准输入中读取输入写入目标文件系统。其中 localsrc 只能是本地文件，dst 只能为 HDFS 文件，且不受 fs.defaultFS 属性影响。

示例：

hadoop fs -put /home/hduser/file/file1.txt /input2

将本地文件复制到 HDFS 目录"/input2"。

hadoop fs -put /home/hduser/file/file1.txt /home/hduser/file/file2.txt /input2

将多个本地文件复制到 HDFS 目录"/input2"。

hadoop fs -put - /input2/file3

从标准输入中读取输入，按 Ctrl+C 组合键退出并保存到"file3"。

（2）get 命令

语法：

hadoop fs -get <src><localdst>

复制 HDFS 文件到本地文件系统，put 命令的逆操作。其中 src 只能为 HDFS 文件，localdst 只能是本地文件，且同样不受 fs.defaultFS 属性影响。

示例：

hadoop fs -get /input2/file1 $HOME/file

将 HDFS 文件"/input2/file1"复制到本地文件系统"$HOME/file"中，$HOME 为 Linux 系统环境变量，代表用户根目录，如 hduser 用户的根目录为"/home/hduser"。

（3）mv 命令

语法：

hadoop fs -mv URI[URI…] <dest>

将文件从源路径移动到目标路径，允许多个源路径，目标路径必须是一个目录。不允许在不同的文件系统间移动文件，也就是说所有路径都必须是同一文件系统 URI 格式。

示例：

hadoop fs -mv /input2/file1.txt /input2/file2.txt /user/hadoop/dir1

将 HDFS 上的 file1.txt、file2.txt 移动到 dir1 中。

（4）cp 命令

语法：

hadoop fs -cp URI [URI…] <dest>

将文件从源路径复制到目标路径，允许多个源路径，目标路径必须是一个目录。不允许在不同的文件系统间复制文件。

示例：

hadoop fs -cp /input2/file1.txt /input2/file2.txt /user/hadoop/dir1

在 HDFS 中复制多个文件到"/user/hadoop/dir1"。

hadoop fs -cp file:///file1.txt file:///file2.txt file:///tmp

在本地文件系统中复制多个文件到目录"file:///tmp"。

5. 删除文件：rm、rmr

（1）rm 命令

语法：

hadoop fs -rm URI [URI…]

删除指定的文件。只删除非空目录和文件。

示例：

hadoop fs -rm /intpu2/file1.txt   #删除非空文件

（2）rmr 命令

语法：

```
hadoop fs -rmr URI [URI…]
```

rm 的递归版本。整个文件夹及子文件夹将全部删除。

示例：

```
hadoop fs -rmr /user/hadoop/dir1        # 递归删除
```

6. 管理命令：test、du、expunge

（1）test 命令

语法：

```
hadoop fs -test -[ 选项 ] URI
```

选项：

- -e：检查文件是否存在。如果存在则返回 0。
- -z：检查文件是否 0 字节。如果是则返回 0。
- -d：检查路径是否为目录，如果是则返回 1，否则返回 0。

示例：

```
hadoop fs -test -e /input2/file3.txt    # 检查文件是否存在
echo $?                                 # "$?" 是 Linux 变量，存储上一条命令的返回值
```

（2）du 命令

语法：

```
hadoop fs -du URI [URI …]
```

显示目录中所有文件的大小。

示例：

```
hadoop fs -du /input2                    # 显示文件的大小，如果是目录则列出所有文件及其大小
hadoop fs -du -s /input2/file1.txt       # 显示文件的大小，如果是目录则统计总大小
```

（3）expunge 命令

语法：

```
hadoop fs -expunge
```

清空回收站。

以上介绍了部分常用命令，更多命令可自行查阅 Hadoop 帮助文档，获取位置：$HADOOP_HOME/share/doc/hadoop/hadoop-project-dist/hadoop-common/FileSystemShell.html。

5.5 MapReduce 基础

为什么使用 MapReduce？对于大量数据的计算，通常采用的处理方法就是并行计

算。这就要求能够将大型而复杂的计算问题分解为各个子任务，并分配到多个计算资源下同时进行计算，其显著特点是耗时小于单个计算资源下的计算。对多数开发人员来说，并行计算还是个陌生、复杂的东西，尤其是涉及到分布式的问题，将会更加棘手。MapReduce 就是一种实现了并行计算的编程模型，它向用户提供接口，屏蔽了并行计算特别是分布式处理的诸多细节，让那些没有多少并行计算经验的开发人员也可以很方便地开发并行应用。

MapReduce 由两个概念合并而来：map（映射）和 reduce（归约）。map 负责把任务分解成多个任务，reduce 负责把分解后多任务的处理结果进行汇总。

5.5.1　MapReduce 概述

我们已经知道，Hadoop 的 MapReduce 框架源自于 Google 的 MapReduce 论文。在 Google 发表论文时，MapReduce 最大成就是重写了 Google 的索引文件系统。现在 MapReduce 被广泛地应用于日志分析、海量数据排序、在海量数据中查找特定模式等场景中。

Hadoop 中的并行应用程序的开发是基于 MapReduce 编程模型的，基于它可以将任务分发到由上千台商用机器组成的集群上，实现 Hadoop 的并行任务处理功能。前面提过，HDFS 和 MapReduce 二者相互作用，共同完成了 Hadoop 分布式集群的主要任务。

5.5.2　MapReduce 架构设计

与 HDFS 架构设计相似，在 Hadoop 中，用于执行 MapReduce 作业的机器也有两个角色：

- JobTracker：是一个 Master 服务，用于作业（Job）的管理和调度工作，一个 Hadoop 集群中只有一台 JobTracker，一般情况应该把它部署在单独的机器上。JobTracker 负责创建、调度作业中的每一个子任务（MapTask 或 ReduceTask）运行于 TaskTracker 上，并监控它们，如果发现有失败的任务就重新运行它。
- TaskTracker：是运行于多个节点上的 Slave 服务，用于执行任务。TaskTracker 需要运行在 HDFS 的 DataNode 上。

基于 JobTracker 和 TaskTracker 的运行架构为 MapReduce V1，在下一代 MapReduce V2 中，V1 架构已被 YARN 替代，关于 YARN 我们稍后会讲解。从学习的难易程序来看，应该先了解 MapReduce V1。不论是 V1 还是 V2，都不会影响我们编写 MapReduce 程序，好比同样是一个 Web 应用，运行在 Tomcat 与 Jetty 下的效果是相同的。由此可见，实际上运行 MapReduce 作业的过程对开发人员是透明的。

5.5.3 MapReduce 编程模型

那么 MapReduce 程序是如何运行的呢？当编写完成 MapReduce 程序，并配置为一个 MapReduce 作业（Job），这里的"作业"可以理解为：为了进行一次分布式计算任务而编写 MapReduce 程序后，将该程序提交到 MapReduce 执行框架中并执行的全过程。当客户端提交 Job 到 JobTracker 后，数据流如图 5.13 所示。

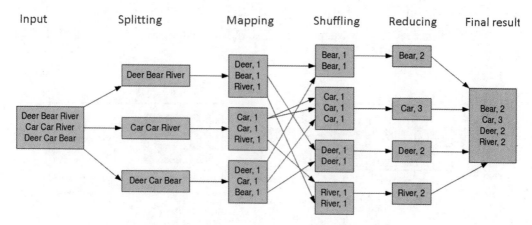

图 5.13　MapReduce 数据流

从图中可以看出，待处理数据从输入到最后输出经过如下五个阶段：

（1）input：由 JobTracker 创建该 Job，并根据 Job 的输入计算输入分片（Input Split）。这里要求待处理的数据集必须可以分解成许多小的数据集，且每一个小的数据集都可以完全并行地进行处理。输入目录中文件的数量决定了分片的数量，又如果对于单个文件超过 HDFS 默认块大小（64MB），将按块大小进行分割。

（2）split：作业调度器获取 Job 的输入分片信息，对输入分片中的记录按照一定规则解析成键值对，"键"（key）是每一行的起始位置，以字节为单位，"值"（value）是本行文本内容。最后每个分片创建一个 MapTask 并分配到某个 TaskTracker。

（3）map：TaskTracker 开始执行 MapTask，处理输入的每一个键值对。如何处理取决于在该阶段的程序代码，处理完成后产生新的键值对，保存在本地。

（4）shuffle：混洗。将 MapTask 的输出有效地作为 ReduceTask 的输入的过程。从图中可以看出该过程会在各 TaskTracker Node 之间进行数据交换，按照 key 进行分组。

（5）reduce：读取 Shuffling 阶段的输出，开始执行 ReduceTask，处理输入的每一个键值对。同样，如何处理取决于该阶段的程序代码。最后输出最终结果。

在 Hadoop 中每个 MapReduce 计算任务都会被初始化为一个 Job。其中主要有两个处理阶段：map 阶段和 reduce 阶段，两个阶段均以键值对 <key,value> 作为输入，然后产生同样以 <key,value> 形式的输出。两个阶段分别对应 map() 和 reduce() 方法，这便是开发人员需要实现的两个最重要的阶段和方法，而其他阶段大多可由系统自动处理。

- 每个 map() 方法负责计算一个输入分片并输出计算结果，由 org.apache.hadoop.mapreduce.Mapper 类提供。每个 MapTask 都会创建一个 Mapper 实例。
- 每个 reduce() 方法负责将多个 map() 的处理结果进行汇总，由 org.apache.hadoop.mapreduce.Reducer 类提供。每个 ReduceTask 都会创建一个 Reducer 实例。

由以上分析可以看出 MapReduce 编程模型所做的工作便是：利用一个输入键值对集合来产生一个输出的键值对集合。

再从总体上来看，一个 MapReduce 作业在 MapReduce 框架中的工作原理如图 5.14 所示。

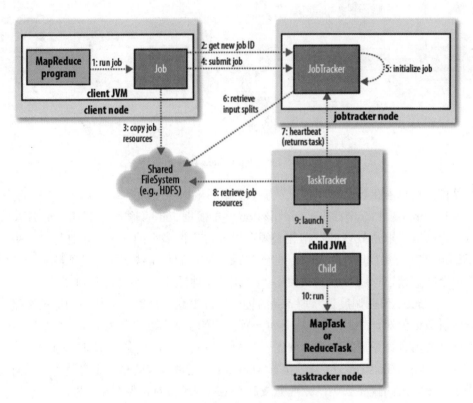

图 5.14　MapReduce 作业的工作原理

从图中可看出，一个 MapReduce 作业的完整运行过程包括 10 个步骤：

（1）编写 MapReduce 程序，包括 Mapper 处理、Reducer 处理以及为执行这些处理而定义的作业，首先将所有这些程序打包后运行作业。

（2）获取作业 ID。

（3）复制作业资源。

（4）提交作业资源。

（5）初始化作业。

（6）获取输入分片。

（7）心跳通信。TaskTracker 运行一个简单的循环来定期发送"心跳"给 JobTracker，表明 TaskTracker 是否还存活，同时也充当两者之间的消息通道。

（8）获取作业资源。

（9）分配任务。

（10）运行任务 MapTask 或 ReduceTask，最后输出 MapReduce 任务处理结果。

除了第一步由开发人员编码实现，其他步骤全部由 Hadoop MapReduce 框架自动执行。换句话说，编写一个 MapReduce 程序，以下三个基本部分是我们需要重点关注的：

- MapTask 程序：Mapper 的实现。
- ReduceTask 程序：Reducer 的实现。
- Job 相关配置。

5.6 下一代 MapReduce 框架 YARN

前面我们都是以旧的 MapReduce V1 框架进行讲解，实际上之前的所有 MapReduce 作业均是运行在 YARN 框架上。在 MapReduce V1 中对于超过 4000 个节点的大型集群，就会开始面临扩展性的瓶颈。主要表现在：

（1）JobTracker 单点瓶颈。MapReduce 中的 JobTracker 负责作业的分发、管理和调度，同时还必须和集群中所有的节点保持"心跳"通信，跟踪机器的运行状态。随着集群的数量和提交 Job 的数量不断增加，导致 JobTracker 的任务量随之增加，最终成为集群的单点瓶颈。

（2）TaskTracker 端，由于作业分配信息过于简单，有可能将多个资源消耗多或运行时间长的 Task 分配到同一个 Node 上，这样会造成作业的等待时间过长。

（3）作业延迟高。在 MapReduce 运行作业之前，需要 TaskTracker 汇报自己的资源运行情况，JobTracker 根据获取的信息分配任务，TaskTracker 获取任务之后再开始运行。这样的结果导致小作业启动时间过长。

（4）编程框架不够灵活。虽然 MapReduce V1 框架允许用户自定义各阶段的对象和处理方法，但 MapReduce V1 还是限制了编程的模式及资源的分配。

基于以上问题，下一代 MapReduce 框架 YARN（Yet Another Resource Negotiator 另一种资源协调者，也称 MapReduce V2）应运而生。

5.6.1 YARN 架构

YARN 将 JobTracker 的职能进行了拆分，从而改善了 MapReduce V1 面临的扩展性瓶颈问题。将 JobTracker 承担的两大块任务：集群资源管理和作业管理进行分离，分别为管理集群上资源使用的资源管理器（ResourceManager）和管理集群上运行任务（MapReduce 作业）生命周期的应用主体（ApplicationMaster），然后 TaskTracker 演化成节点管理器（NodeManager），如图 5.15 所示。

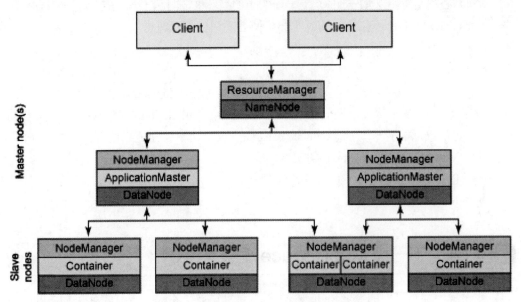

图 5.15　YARN 架构

　　YARN 仍然是 Master/Slave 结构，全局的 ResourceManager 和局部的 NodeManager 组成了数据计算框架，ApplicationMaster 负责与 ResourceManager 通信获取资源并与 NodeManager 配合完成节点的 Task 任务。下面列出了各实体角色的职责：

　　（1）资源管理器：包括两个功能组件调度器和应用管理器。调度器仅负责协调集群上计算资源的分配，不负责监控各个应用的执行情况。应用管理器负责接收作业，协商获取第一个资源容器用于启动作业所属的应用主体并监控应用主体的存在情况。

　　（2）节点管理器：负责启动和监视集群中机器上的计算资源容器（Container）。

　　（3）应用主体：应用主体与应用一一对应，负责协调运行 MapReduce 作业的任务，它和 MapReduce 任务都在资源容器中运行。

　　（4）资源容器：对节点自身内存、CPU、磁盘、网络带宽等资源的抽象封装，由资源管理器分配并由节点管理器进行管理。主要职责是运行、保存或传输应用主体提交的作业或需要存储和传输的数据。

5.6.2　YARN 配置文件

　　前面提过，基于 MapReduce V1 编写的程序无需修改也可以运行在 YARN 中。启动 YARN 需要单独配置，共涉及到如下文件：

　　（1）yarn-env.sh：加入 JDK 路径。

　　（2）mapred-site.xml：指定 mapreduce.framework.name 为 YARN。

　　（3）yarn-site.xml：YARN 具体配置信息，详细如表 5-5 所示。

表 5-5　YARN 主要配置属性

属性	默认值	说明
yarn.reourcemanager.address	hostname:8032	ResourceManager 对客户端暴露的地址。客户端通过该地址向 ResourceManager 提交、终止应用程序
yarn.resourcemanager.scheduler.address.	hostname:8030	ResourceManage 对 ApplicationMaster 暴露的地址。ApplicationMaster 通过该地址向 ResourceManager 申请、释放资源
yarn.resourcemanager.resource-tracker.address	hostname:8031	ResourceManage 对 NodeManager 暴露的地址。NodeManager 通过该地址向 ResourceManager 汇报心跳，领取任务
yarn.resourcemanager.admin.address	hostname:8033	ResourceManage 对管理员暴露的地址。管理员通过该地址向 ResourceManager 发送管理命令
yarn.resourcemanager.webapp.address	hostname:8088	ResourceManage 对外 Web 访问地址。用户可通过该地浏览器中查看集群各类信息

另外 YARN 需要单独启动：sbin/start-yarn.sh，在 NameNode 上使用"jps"查看到"ResourceManager"进程表示已成功运行，相应地在 DataNode 上会出现"NodeManager"进程。

5.6.3　YARN 作业执行流程

在 YARN 中执行流程如图 5.16 所示。

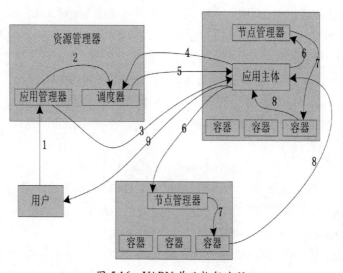

图 5.16　YARN 作业执行流程

（1）MapReduce 框架接收用户提交的作业，并为其分配一个新的应用 ID，并将应用的定义打包上传到 HDFS 上用户的应用缓存目录中，然后提交此应用给应用管理器。

（2）应用管理器同调度器协商获取运行应用主体所需的第一个资源容器。

（3）应用管理器在获取的资源容器上执行应用主体。

（4）应用主体计算应用所需资源，并发送资源请求到调度器。

（5）调度器根据自身统计的可用资源状态和应用主体的资源请求，分配合适的资源容器给应用主体。

（6）应用主体与所分配容器的节点管理器通信，提交作业情况和资源使用说明。

（7）节点管理器启用容器并运行任务。

（8）应用主体监控容器上任务的执行情况。

（9）应用主体反馈作业的执行状态信息和完成状态。

5.6.4　YARN 优势

（1）分散了 JobTracker 的任务，资源管理器和应用主体各司其职，解决了 JobTracker 的瓶颈问题，提高了集群的扩展性。

（2）YARN 中应用主体是一个用户可自定义的部分，用户可以针对编程模型编写自己的应用主体程序，扩展了 YARN 的适用范围。

（3）集群资源统一组织成资源容器，提高了集群资源的利用率。

本章总结

- Hadoop 包含了 HDFS、MapReduce/YARN、Hive、Pig、HBase、ZooKeeper 等子项目，本门课程将学习 HDFS、MapReduce 和 HBase。
- Hadoop 的版本分为两代，主要区别是新一代的 MapReduce 称为 YARN。
- Hadoop 的核心是 HDFS 和 MapReduce。HDFS 完成分布式存储，MapReduce 解决分布式计算。
- Hadoop 的运行环境包括：Linux 系统、JDK、SSH 及 Hadoop 自身。
- Hadoop 可有三种运行方式：单机、伪分布式及完全分布式。前两者往往用于测试环境。在完全分布式环境中，不同的系统对于主机的角色划分是不同的，但都遵从主从架构：
    - 在 HDFS 中的节点分为 NameNode 和 DataNode。
    - 在 MapReduce 中的节点分为 JobTracker 和 TaskTracker。
- Hadoop 主要配置文件包括 hadoop-env.sh、yarn-env.sh、slaves、core-site.xml、hdfs-site.xml、mapred-site.xml、yarn-site.xml。
- Hadoop HDFS 采用 Master/Slave 架构，由一个 NameNode 和多个 DataNode 组成。NameNode 作为主服务器负责管理文件系统的命名空间和客户端对文件的访问操作，DataNode 负责管理存储的数据。HDFS 中的文件被分成文件块存储在不同的 DataNode 中，每个文件块的默认大小为 64MB。

- 通过 HDFS 的 Web 界面来监控 HDFS 的基本信息，包括集群启动时间、版本号、NameNode 基本信息与所有 DataNode 信息等，另外还提供浏览文件系统、日志等功能。
- HDFS 的基本命令行操作：mkdir、ls、cat、put、get、mv、cp、rm、rmr、test、du、expunge。
- MapReduce 是在 Hadoop 中进行并行应用程序开发的编程模型，MapReduce 程序包含两个处理核心：Mapper 和 Reducer。
- MapReduce 作业主要配置项：输入 / 输出（格式、路径、key/value 类型）、Mapper 类、Reducer 类。
- MapReduce 数据流：input → split → map → shuffle → reduce。
- MapReduce 应用程序在 MapReduce V1 和 YARN 中均可运行。推荐配置使用 YARN 框架，它提高了集群的扩展性、适用范围和集群资源的利用率。

本章作业

1. 简述 Hadoop 集群安装的步骤。
2. 运行 MapReduce 程序的步骤是什么？请详细说明并写出必要的命令。
3. 用课工场 APP 扫一扫完成在线测试，快来挑战吧！

随手笔记

第 6 章

HBase 数据库

技能目标

- 了解 HBase 体系结构
- 理解 HBase 数据模型
- 掌握 HBase 的安装
- 会使用 HBase Shell 操纵 HBase

本章导读

　　本章我们将学习 Hadoop 的数据库 HBase，包括 HBase 的体系结构、数据模型、安装及其操作。通过这一章的学习，将了解什么是面向列的数据库，并掌握 HBase 数据库的分布式架构、数据模型以及其安装和基本使用方法。

知识服务

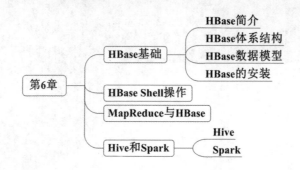

6.1 HBase 基础

6.1.1 HBase 简介

HBase 是一个基于 HDFS 的面向列的分布式数据库，源于 Google 的 BigTable 论文。前面提过，HDFS 是基于流式数据访问，对于低时间延迟的数据访问并不适合在 HDFS 上运行。所以如果需要实时地随机访问超大规模的数据集，使用 HBase 则是更好的选择。

HBase 是数据库，但并不是传统的关系型数据库，如 Oracle、SQL Server、MySQL 等，这些数据库本身并不是为了可扩展的分布式处理而设计。首先 HBase 不支持关系型数据库的 SQL，并且不使用以行存储的关系型结构存储数据，而是以键值对的方式按列存储，由此我们认为它是非关系型数据库 NoSQL（Not Only SQL）中的一个重要代表。NoSQL 的概念在 2009 年被提出，目前并没有明确的范围和定义，主要特点是通常用于大规模数据的存储、没有预定义的模式（如表结构）、表和表之间没有复杂的关系。总体上可将 NoSQL 数据库分为以下四类：

- 基于列存储的类型
- 基于文档存储的类型
- 基于键值对存储的类型
- 基于图形数据存储的类型

通常人们将 HBase 归为基于列存储类型这一类，在 NoSQL 领域，HBase 本身不是最优秀的，但得益于与 Hadoop 的整合，给它带来了更广阔的发展空间。HBase 本质上只有插入操作，更新和删除都是使用插入方式完成，这是由底层 HDFS 流式访问特性（一次写入、多次读取）决定的。所以在更新时总是插入一个带时间戳的新行，而删除时插入一个带有删除标记的新行。每一次的插入都有一个时间戳标记，每次都是一个新的版本，HBase 会保留一定数量的版本，这个值是可以设定的。如果在查询时提供时间戳则返回距离该时间最近的版本，否则返回离现在最近的版本。

6.1.2 HBase 体系结构

HBase 的服务器体系结构同样是 Master/Slaves 的主从服务器结构，它由一个 HMaster 服务器和多个 HRegionServer 服务器构成，而所有这些服务器都是通过 ZooKeeper 来进行协调并处理各服务器运行期间可能遇到的错误。HMaster 负责管理所有的 HRegionServer，各 HRegionServer 负责存储许多 HRegion，每一个 HRegion 是对 HBase 逻辑表的分块。如图 6.1 所示，图中给出了 HBase 集群中的所有成员。

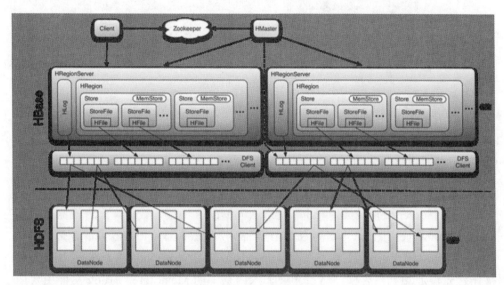

图 6.1　HBase 体系结构

下面针对图 6.1 中的主要实体分别进行介绍。

1．HRegion

HBase 使用表（Table）存储数据集，表由行和列组成，这与关系型数据库类似。但是在 HBase 中，当表的大小超过设定值时，HBase 会自动将表划分为不同的区域（Region），每个区域称为 HRegion，它是 HBase 集群上分布式存储和负载均衡的最小单位，在这点上表和 HRegion 类似于 HDFS 中文件与文件块的概念。一个 HRegion 中保存一个表中一段连续的数据，通过表名和主键范围（开始主键～结束主键）来区分每一个 HRegion。一开始，一个表只有一个 HRegion，随着 HRegion 开始变大，直到超出设定的大小阈值，便会在某行的边界上把表分成两个大小基本相同的 HRegion，称为 HRegion 分裂，如图 6.2 所示。

每个 HRegion 由多个 HStore 组成，每个 HStore 对应表中一个列族（Column Family）的存储，列族在后面还有详细介绍。HStore 由两部分组成：MemStore 和 StoreFile，用户写入的数据首先放入 MemStore，当 MemStore 满了以后再刷入（flush）StoreFile。StoreFile 是 HBase 中的最小存储单元，底层最终由 HFile 实现，而 HFile 是键值对数据的存储格式，实质是 HDFS 的二进制格式文件。

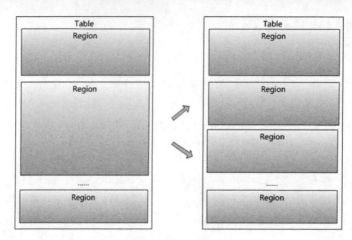

图 6.2　HRegion 分裂

HBase 中不能直接更新和删除数据，所有的数据均通过追加的方式进行更新。当 StoreFile 的数量超过设定的阈值将触发合并操作，将多个 StoreFile 合并为一个 StoreFile，此时进行数据的更新和删除。

2. HRegionServer

HRegionServer 负责响应用户 I/O 请求，向 HDFS 中读写数据，一台机器上只运行一个 HRegionServer。HRegionServer 包含两部分：HLog 部分和 HRegion 部分。

其中 HLog 用于存储数据日志，实质是 HDFS 的 Sequence File。到达 HRegion 的写操作首先被追加到日志中，然后才被加入内存中的 MemStore。HLog 文件主要用于故障恢复。例如某台 HRegionServer 发生故障，那么它所维护的 HRegion 会被重新分配到新的机器上，新的 HRegionServer 在加载 HRegion 的时候可以通过 HLog 对数据进行恢复。

HRegion 部分由多个 HRegion 组成，每个 HRegion 对应了表中的一个分块，并且每一个 HRegion 只会被一个 HRegionServer 管理。

3. HMaster

每台 HRegionServer 都会和 HMaster 服务器通信，HMaster 的主要任务就是告诉每个 HRegionServer 它要维护哪些 HRegion。

在 HBase 中可以启动多个 HMaster，通过 ZooKeeper 的 Master 选举机制来保证系统中总有一个 Master 在运行。HMaster 的具体功能包括：

- 管理用户对表的增、删、改、查操作。
- 管理 HRegionServer 的负载均衡，调整 HRegion 分布。
- 在 HRegion 分裂后，负责新 HRegion 的分配。
- 在 HRegionServer 停机后，负责失效 HRegionServer 上的 HRegion 迁移。

4. ZooKeeper

ZooKeeper 存储的是 HBase 中的 ROOT 表和 META 表的位置，这是 HBase 中两张

特殊的表，称为根数据表（ROOT）和元数据表（META）。META 表记录普通用户表的 HRegion 标识符信息，每个 HRegion 的标识符为：表名 + 开始主键 + 唯一 ID。随着用户表的 HRegion 的分裂，META 表的信息也会增长，并且可能还会被分割为几个 HRegion，此时可以用一个 ROOT 表来保存 META 的 HRegion 信息，而 ROOT 表是不能被分割的，也就是 ROOT 表只有一个 HRegion。那么客户端（Client）在访问用户数据前需要首先访问 ZooKeeper，然后访问 ROOT 表，接着访问 META 表，最后才能找到用户数据的位置进行访问，如图 6.3 所示。

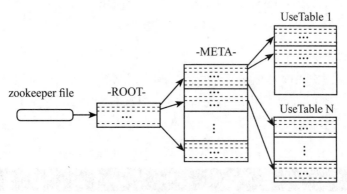

图 6.3　ROOT 表与 META 表

此外 ZooKeeper 还负责监控各个机器的状态，之前各机器需要在 ZooKeeper 中注册一个实例。当某台 HRegionServer 发生故障时，通知 HMaster 进行 HRegion 迁移；若 HMaster 发生故障，ZooKeeper 负责恢复 HMaster，并且保证同时有且只有一台 HMaster 运行。

6.1.3　HBase 数据模型

1. 数据模型

在 HBase 中，数据以表的方式存储。具体数据模型中涉及到的术语解释如下：

- 表（Table）：是一个稀疏表（不存储值为 NULL 的数据），表的索引是行关键字、列关键字和时间戳。
- 行关键字（Row Key）：行的主键，唯一标识一行数据，也称行键。表中的行根据行键进行字典排序，所有对表的访问都要通过表的行键，在创建表时，行键不用、也不能预先定义，而在对表数据进行操作时必须指定行键，行键在添加数据时首次被确定。
- 列族（Column Family）：行中的列被分为"列族"。同一个列族的所有成员具有相同的列族前缀。例如 "course:math" 和 "course:art" 都是列族 "course" 的成员。一个表的列族必须在创建表时预先定义，列族名称不能包含 ASCII 控制字符（ASCII 码在 0 ～ 31 间外加 127）和冒号（:）。

- 列关键字（Column Key）：也称列、列键。格式为："<family>:<qualifier>"，其中 family 是列族名，用于表示列族前缀；qualifier 是列族修饰符，表示列族中的一个成员，列族成员可以在随后使用时按需加入，也就是只要列族预先存在，我们随时可以把列族成员添加到列族中去。列族修饰符可以是任意字节。
- 存储单元格（Cell）：在 HBase 中，值是作为一个单元保存在系统中的，要定位一个单元，需要使用"行键 + 列键 + 时间戳"三个要素。
- 时间戳（Timestamp）：插入单元格时的时间戳，默认作为单元格的版本号。

下面结合 HBase 的概念视图再来体会这些术语。

2. 概念视图

在关系型数据库中只能通过表的主键或唯一字段定位到某一条数据，例如使用典型的关系表的结构来描述学生成绩表 scores，如表 6-1 所示，其中主要字符依次为姓名（name）、年级（grade）、数学成绩（math）、艺术成绩（art），主键为 name。

表 6-1　典型的关系表——学生成绩表 scores

name	grade	math	art
json	2	57	87
tom	1	89	80

由于主键唯一标识了一行记录，所于我们很容易按姓名查询到某位同学的所有成绩。但是请思考如下问题。

（1）如果现在新加一门课程，能够在不修改表结构的情况下去保存新的课程成绩吗？

（2）如果某同学数学课程参加了补考，那么两次的考试成绩都能够保存下来吗？

（3）如果某同学只考试了一门课程而其他课程都没有成绩，是否我们可以只保存有成绩的课程而节省存储空间呢？

对于问题（1）和（2），在不修改表结构的情况下是不能够实现的，即使通过修改表结构实现，也不能保证后续的需求不会再发生变化。而在问题（3）中，按表结构的字段类型定义，一条记录的某个字段无论是否为 NULL 都会占用存储空间，那么我们将不能有选择地保存数据来节省存储空间。

很明显，上面的需求在实际运用中经常出现，而 HBase 则可以完美地解决这些问题。我们将 scores 表转为在 HBase 中的概念视图，如表 6-2 所示。

表中的列包含了行键、时间戳和两个列族（grade、course），行包含了"jason"和"tom"两行数据。每次对表操作都必须指定行键和列键，每次操作增加一条数据，每一条数据对应一个时间戳，从上往下按倒序排列，该时间戳自动生成，不必用户管理。每次只针对一个列键操作，例如在 t3 时刻，用户指定"tom"的"math"成绩为"89"，类似操作：先找到行键"tom"，然后指定列键进行赋值：course:math=89，其中"course:math"为列键，"course"为列族名而"math"为列族修饰符，最后将"89"作为列键值赋给列键，t3 时刻的时间自动插入到时间戳这列中。

表 6-2　scores 表在 HBase 中的概念视图

行键（name）	时间戳	列族（grade）		列族（course）	
		列关键字	值	列关键字	值（单元格）
jason	t6			course:math	57
	t5			course:art	87
	t4	grade:	2		
tom	t3			course:math	89
	t2			course:art	80
	t1	grade:	1		

又如在 t4 时刻，用户指定"jason"的"grade"为"2"。按照前面分析应类似如下操作：先找到行键"jason"，然后指定列键并赋值：grade:=2，请注意这里没有给出列键的列族修饰符，即列族修饰符为空字符串。这样是允许的，因为前面提过列族修饰符可以是任意字符。

现在在 HBase 中回答前面的三个问题：

（1）如果现在为学生 jason 新增英语成绩，那么指定行键："jason"，列键："cource:english"，以及列键值（英语成绩）即可。

（2）如果学生 jason 参加了数学补考，那么指定行键："jason"，列键："cource:math"，以及列键值（补考成绩）即可。

（3）前面提过 HBase 是基于稀疏存储设计，在概念视图中发现存在很多空白项，这些空白项并不会被实际存储，总之是有数据就存储无数据则忽略，通过表的物理视图可以更好地体会这一点。

3. 物理视图

通过概念视图，有助于我们从逻辑上理解 HBase 的数据结构，但在实际存储时，是按照列族来存储的，一个新的列键可以随时加入到已存在的列族中，这也是列族必须在创建表时预先定义的原因。由表 6-2 中的概念视图对应的物理视图如表 6-3 和表 6-4 所示。

表 6-3　scores 表在 HBase 中的物理视图（1）

行键（name）	时间戳	列族（grade）	
		列关键字	值
jason	t4	grade:	2
tom	t1	grade:	1

表 6-4 scores 表在 HBase 中的物理视图（2）

行键（name）	时间戳	列族（course）	
		列关键字	值
jason	t6	course:math	57
	t5	course:art	87
tom	t3	course:math	89
	t2	course:art	80

HBase 就是这样一个基于列模式的映射数据库，它只能表示简单的键－值的映射关系。与关系型数据库相比，它有如下特点：

- 数据类型：HBase 只有简单的字符串类型，它只保存字符串。而关系型数据库有丰富的类型选择和存储方式。
- 数据操作：HBase 只有简单的插入、查询、删除、清空等操作，表和表之间是分离的，没有复杂的表和表之间的关系，所以不能、也没有必要实现表和表之间的关联操作。而关系型数据库有多种连接操作。
- 存储模式：HBase 是基于列存储的，每个列族都由几个文件保存，不同列族的文件是分离的。关系型数据库是基于表格结构和行模式存储的。
- 数据维护：HBase 的更新操作实际上是插入了新的数据，它的旧版本依然会保留，而不是关系型数据库的替换修改。
- 可伸缩性：HBase 这类分布式数据库就是为了这个目的而开发出来的，所以它能够轻松地增加或减少硬件数量，并且对错误的兼容性比较高。而关系型数据库通常需要增加中间层才能实现类似的功能。

6.1.4 HBase 的安装

安装 HBase 与 HDFS、MapReduce 不同，HBase 需要单独安装，首先下载压缩包，我们提供的是 hbase-1.0.2 版本。然后将其解压到 NameNode 节点（node1）上：

tar xvf hbase-1.0.2-bin.tar.gz -C /home/hduser

图 6.4 列出了 HBase 的目录结构。

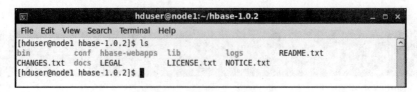

图 6.4 HBase 目录结构

下面是其中各目录的说明：
- bin：包含了所有可执行命令与脚本。
- conf：配置文件目录。
- hbase-webapps：存储 Web 应用的目录，里面应用主要用于查看 HBase 运行状态。默认访问地址 http://Master:16010，其中 Master 为 HBase Master 服务器地址。
- lib：jar 文件目录，包括第三方依赖与 Hadoop 相关 jar 文件。其中 Hadoop 相关 jar 文件版本最好能与实际运行的 Hadoop 版本一致，保证稳定运行。
- logs：日志目录。

按照惯例，HBase 的 conf 目录下也提供了 hbase-site.xml 文件进行自定义配置。用以覆盖默认配置文件 hbase-default.xml，其位于 lib/hbase-common-1.0.2.jar 中。按照 hbase-site.xml 不同的配置方式，使得 HBase 分别运行在单机、伪分布式和完全分布模式下，其中运行分布式 HBase 需要以下条件：
- JDK 环境
- SSH 免密码登录
- Hadoop 环境

1. 单机模式

解压后即可在单机模式下运行，在此模式下只需要在 hbase-site.xml 中指定 HBase 的文件存储目录即可，如下所示：

```
<configuration>
    <property>
            <name>hbase.rootdir</name>
            <value>file:///home/hduser/hbase</value>
    </property>
</configuration>
```

其中 hbase.rootdir 指定了 HBase 的数据存储目录，注意这是 Linux 系统的文件目录。运行下面命令即可启动 HBase：

bin/start-hbase.sh

启动成功后如图 6.5 所示。可以看到，在单机模式下的 HBase 运行进程仅有 HMaster 进程。

```
[hduser@node1 hbase-1.0.2]$ ./bin/start-hbase.sh
starting master, logging to /home/hduser/hbase-1.0.2/bin/../logs/hbase-hduser-master-node1.out
[hduser@node1 hbase-1.0.2]$ jps
23733 HMaster
23801 Jps
[hduser@node1 hbase-1.0.2]$
```

图 6.5　启动 HBase

启动后 HBase 自动创建 hbase.rootdir 目录，其中的文件数据如图 6.6 所示。

```
[hduser@node1 hbase-1.0.2]$ ls /home/hduser/hbase
data  hbase.id  hbase.version  oldWALs  WALs
[hduser@node1 hbase-1.0.2]$
```

图 6.6　HBase 数据存储目录

停止 HBase 使用如下命令：

bin/stop-hbase.sh

如图 6.7 所示。

```
[hduser@node1 hbase-1.0.2]$ ./bin/stop-hbase.sh
stopping hbase...................
[hduser@node1 hbase-1.0.2]$ jps
24308 Jps
[hduser@node1 hbase-1.0.2]$
```

图 6.7　停止 HBase

2. 伪分布式模式

在伪分布式模式下，HBase 只在单个节点上运行，这和单机模式一样，但是其数据文件可以存储在 HDFS 分布式存储系统中。配置伪分布模式我们只需要在 hbase-site.xml 中将 hbase.rootdir 的值更换为 HDFS 文件系统即可。修改为如下配置：

```
<configuration>
    <property>
        <name>hbase.rootdir</name>
        <value>hdfs://node1:9000/hbase</value>
    </property>
</configuration>
```

使用 HDFS 替换本地文件系统后，必须首先启动 HDFS，如图 6.8 所示。

```
[hduser@node1 hbase-1.0.2]$ ~/hadoop/sbin/start-dfs.sh
Starting namenodes on [node1]
node1: starting namenode, logging to /home/hduser/hadoop/logs/hadoop-hduser-namenode-node1.out
node2: starting datanode, logging to /home/hduser/hadoop/logs/hadoop-hduser-datanode-node2.out
node3: starting datanode, logging to /home/hduser/hadoop/logs/hadoop-hduser-datanode-node3.out
Starting secondary namenodes [node1]
node1: starting secondarynamenode, logging to /home/hduser/hadoop/logs/hadoop-hduser-secondarynamenode-node1.out
[hduser@node1 hbase-1.0.2]$ ./bin/start-hbase.sh
starting master, logging to /home/hduser/hbase-1.0.2/bin/../logs/hbase-hduser-master-node1.out
[hduser@node1 hbase-1.0.2]$ jps
25610 HMaster
25402 SecondaryNameNode
25891 Jps
25220 NameNode
[hduser@node1 hbase-1.0.2]$
```

图 6.8　伪分布模式启动 HBase

从图 6.8 中可以看出，进程列表主要包含 Hadoop 的相关进程和 HBase 的 HMaster 进程，对 HBase 来说还是只有一个进程，这和单机模式并无差异。但在此模式下 HBase 数据存储目录位于 HDFS 中，如图 6.9 所示。

```
[hduser@node1 hbase-1.0.2]$ ~/hadoop/bin/hadoop fs -ls /hbase
Found 7 items
drwxr-xr-x   - hduser supergroup          0 2016-07-04 14:38 /hbase/.tmp
drwxr-xr-x   - hduser supergroup          0 2016-07-04 14:38 /hbase/WALs
drwxr-xr-x   - hduser supergroup          0 2016-07-04 13:02 /hbase/corrupt
drwxr-xr-x   - hduser supergroup          0 2016-07-04 11:05 /hbase/data
-rw-r--r--   2 hduser supergroup         42 2016-07-04 11:05 /hbase/hbase.id
-rw-r--r--   2 hduser supergroup          7 2016-07-04 11:05 /hbase/hbase.version
drwxr-xr-x   - hduser supergroup          0 2016-07-04 14:39 /hbase/oldWALs
[hduser@node1 hbase-1.0.2]$
```

图 6.9　伪分布模式下的 HBase 数据存储目录

3. 完全分布式模式

完全分布式与伪分布式的差别在于 HBase 会运行在多个节点上。通常是将 HBase 的 HMaster 运行在 HDFS 的 NameNode 上，而 HRegionServer 运行在 HDFS DataNode 上。继续前面的实验环境，各节点主机角色分配如表 6-5 所示。

表 6-5　主机名与对应的 IP 地址、角色

主机名	IP 地址	所分配的角色
node1	192.168.70.130	Master，NameNode，JobTracker，HMaster
node2	192.168.70.131	Slave，DataNode，TaskTracker，HRegionServer
node3	192.168.70.132	Slave，DataNode，TaskTracker，HRegionServer

在后续的实验中 HBase 均采用完全分布式模式运行，在此模式下需要在 conf 目录下配置三个文件 hbase-site.xml、hbase-env.sh 和 regionservers。我们先在 node1 上进行配置，随后将整个 HBase 安装目录复制到其他节点上。

在配置前先做一些必要的清理工作：

- 将 HDFS 中已经存在的"hdfs://node1:9000/hbase"目录删除（按前面配置示例，配置过伪分布式模式运行，已使用过该目录），如果没有请跳过。
- 检查并同步所有节点机（node1、node2、node3）的时钟，并且各节点与 HBase 的 HMaster 节点（node1）时钟误差不能大于 30 秒。

下面针对三个配置文件分别说明。

（1）hbase-site.xml 文件

完整配置如下：

```
<configuration>
    <property>
        <name>hbase.rootdir</name>
        <value>hdfs://node1:9000/hbase</value>
        <description> 配置 HRegionServer 的数据库存储目录 </description>
    </property>
```

```xml
<property>
    <name>hbase.cluster.distributed</name>
    <value>true</value>
    <description> 配置 HBase 为完全分布式 </description>
</property>
<property>
    <name>hbase.master</name>
    <value>node1:60000</value>
    <description> 配置 HMaster 的地址 </description>
</property>
<property>
    <name>hbase.zookeeper.quorum</name>
    <value>node1,node2,node3</value>
    <description> 配置 ZooKeeper 集群服务器的位置 </description>
</property>
</configuration>
```

其中主要参数说明如下：

- hbase.cluster.distributed：默认为 false，即单机或伪分布式运行。这里设置为 true，表示在完全分布式模式运行。
- hbase.master：指定 HBase 的 HMaster 服务器地址、端口。
- hbase.zookeeper.quorum：指出了 ZooKeeper 集群中各服务器位置。也就是将哪些节点加入到 ZooKeeper 进行协调管理，推荐为奇数个服务器。

（2）hbase-env.sh 文件

此文件用来配置全局的 HBase 集群系统的特性，每一台机器都可以通过该文件来了解全局的 HBase 的某些特性。需要在文件末尾增加以下环境变量：

```
export JAVA_HOME=/usr/java/jdk1.7.0_67
export HADOOP_HOME=/home/hduser/hadoop
export HBASE_HOME=/home/hduser/hbase-1.0.2
export HBASE_MANAGES_ZK=true
```

前三个环境变量分别代表 Java、Hadoop、HBase 安装目录。完全分布式的 HBase 集群需要 ZooKeeper 实例的运行，那么最后一个环境变量 HBASE_MANAGES_ZK 表示 HBase 是否使用内置的 ZooKeeper 实例，默认为 true，当在 hbase-site.xml 文件中配置了 hbase.zookeeper.quorum 属性后，系统会使用该属性所指定的 ZooKeeper 集群服务器列表，在启动 HBase 时，HBase 将把 ZooKeeper 作为自身的一部分运行，其对应进程为"HQuorumPeer"，关闭 HBase 时其内置 ZooKeeper 实例也一起关闭。如果 HBASE_MANAGES_ZK 为 false，表示不会使用内置 ZooKeeper 实例，也就是内置 ZooKeeper 不会随 HBase 启动，而需要用户在指定机器上独立安装配置 ZooKeeper 实例，同样使用 hbase.zookeeper.quorum 属性指定这些机器，并且在启动 HBase 之前必须手动启动这些机器的 ZooKeeper。

> **说明**
>
> 为了方便讲解，我们使用 HBase 内置 ZooKeeper 实例，关于 ZooKeeper 的单独安装与配置不在本课程讨论范围内。如果需要，同学们可以自行查阅相关资料。

（3）regionservers 文件

该文件列出了所有 HRegionServer 节点，配置方式与 Hadoop 的 slaves 文件类似，每一行指定一台机器。当 HBase 启动、关闭时会把此文件中列出的所有机器同时启动、关闭。按表 6-5 的各机器角色分配，我们将 node2、node3 作为 HRegionServer。故 regionservers 文件中内容如下：

```
node2
node3
```

> **注意**
>
> regionservers 文件不包含 node1，因为 node1 已在 hbase-site.xml 中被指定为 HMaster 服务器，通常不会将 HMaster 和 HRegionServer 服务器运行在一个节点上。

在机器 node1 上配置完成上面三个文件后，HBase 基本的完全分布式模式配置便已完成。同 Hadoop 分布式安装类似，还需要将 HBase 所在目录如 "/home/hduser/hbase-1.0.2" 分别复制到 node2、node3，使得各个节点上都能运行 HBase 来构建 HBase 集群。

在 node1 上运行下面命令：

```
scp -r /home/hduser/hbase-1.0.2 hduser@node2:/home/hduser/
scp -r /home/hduser/hbase-1.0.2 hduser@node3:/home/hduser/
```

启动 HBase 如图 6.10 所示。HBase 会先启动 ZooKeeper，再启动所有 HMaster 和 HRegionServer，启动成功后注意 node1 上的 Java 进程增加了两个："HQuorumPeer" 和 "HMaster"，分别为 ZooKeeper 进程和 HBase 进程。

再看 node2 的 java 进程，这里多了两个进程 "HQuorumPeer" 和 "HRegionServer"，同样分别为 ZooKeeper 进程和 HBase 进程，如图 6.11 所示。另外在 node3 中的 Java 进程应该和 node2 一样。

在启动 HBase 后，通过命令 "hbase shell" 进入 HBase Shell，然后使用 HBase Shell 命令 "status" 可在 HBase Shell 中查看 HBase 的运行状态。如图 6.12 所示，表示当前共有 2 个 HRegionServer 正在正常运行。

```
[hduser@node1 hbase-1.0.2]$ bin/start-hbase.sh
node2: starting zookeeper, logging to /home/hduser/hbase-1.0.2/logs/hbase-hduser-zook
eeper-node2.out
node1: starting zookeeper, logging to /home/hduser/hbase-1.0.2/logs/hbase-hduser-zook
eeper-node1.out
node3: starting zookeeper, logging to /home/hduser/hbase-1.0.2/logs/hbase-hduser-zook
eeper-node3.out
starting master, logging to /home/hduser/hbase-1.0.2/logs/hbase-hduser-master-node1.o
ut
node3: starting regionserver, logging to /home/hduser/hbase-1.0.2/logs/hbase-hduser-r
egionserver-node3.out
node2: starting regionserver, logging to /home/hduser/hbase-1.0.2/logs/hbase-hduser-r
egionserver-node2.out
[hduser@node1 hbase-1.0.2]$ jps
27055 HQuorumPeer
27273 Jps
25402 SecondaryNameNode
27150 HMaster
25220 NameNode
[hduser@node1 hbase-1.0.2]$
```

图 6.10 完全分布式模式 HBase 的启动

```
[hduser@node2 ~]$ jps
69751 HQuorumPeer
69403 DataNode
70002 Jps
69815 HRegionServer
[hduser@node2 ~]$
```

图 6.11 node2 的进程

```
[hduser@node1 hbase-1.0.2]$ ./bin/hbase shell
SLF4J: Class path contains multiple SLF4J bindings.
SLF4J: Found binding in [jar:file:/home/hduser/hbase-1.0.2/lib/slf4j-log4j12-1.7.7.ja
r!/org/slf4j/impl/StaticLoggerBinder.class]
SLF4J: Found binding in [jar:file:/home/hduser/hadoop/share/hadoop/common/lib/slf4j-l
og4j12-1.7.5.jar!/org/slf4j/impl/StaticLoggerBinder.class]
SLF4J: See http://www.slf4j.org/codes.html#multiple_bindings for an explanation.
SLF4J: Actual binding is of type [org.slf4j.impl.Log4jLoggerFactory]
HBase Shell; enter 'help<RETURN>' for list of supported commands.
Type "exit<RETURN>" to leave the HBase Shell
Version 1.0.2, r76745a2cbffe08b812be16e0e19e637a23a923c5, Tue Aug 25 15:59:49 PDT 201
5

hbase(main):001:0> status
2 servers, 0 dead, 1.0000 average load

hbase(main):002:0>
```

图 6.12 查看 HBase 的所有 HRegionServer 状态

> **提示**
>
> 如输入命令"hbase shell"提示"bash: hbase: command not found"，此时将 HBase 的 bin 目录加入到系统环境变量 PATH 中即可。方法如下：
> (1) 打开文件：sudo gedit /etc/profile。
> (2) 在打开文件中增加一行："export PATH=/home/hduser/hbase-1.0.2/bin:$PATH"，随后保存退出。
> (3) 使配置生效：source /etc/profile。

还可以通过 HMaster 节点的 16010 端口查看 HBase 的运行状态，如 http://node1:16010，页面输出如图 6.13 所示。

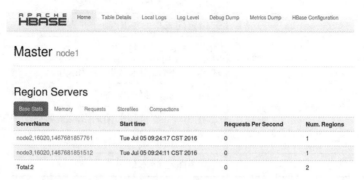

图 6.13　使用 Web 方式查看 HBase 运行状态

最后退出 HBase Shell 使用命令"exit"即可。

6.2　HBase Shell 操作

前面我们已经接触了两个 HBase Shell 命令：status 和 exit。HBase 命令很多，其中又分为几个组，输入：help"cmd" 可查看所有分组及其所包含的命令，如果需要了解具体命令的用法将参数"cmd"换成具体命令即可，如：help"status"。针对基本的使用来说，掌握下面常用的命令即可，如表 6-6 所示。

表 6-6　HBase Shell 常用命令

HBase Shell 命令	说明
alter	修改表的列族
count	统计表中行的数量，一个行键为一行
create	创建表
describe	显示表的详细信息
delete	删除指定对象的值
deleteall	删除指定行的所有元素值
disable	使表无效
drop	删除表
enable	使表有效
exists	测试表是否存在
exit	退出 HBase Shell
get	获取行或单元的值
incr	增加指定表、行或列的值

HBase Shell 命令	说明
list	列出 HBase 中存在的所有表
put	向指定的表单元添加值
tools	列出 HBase 所支持的工具
scan	通过对表的扫描来获取对应的值
status	返回 HBase 集群的状态信息
shutdown	关闭 HBase 集群（关闭后必须重新启动 HBase）
truncate	重新创建指定表
version	返回 HBase 版本信息

现在使用命令将表 6-2 所示的概念视图 scores 表保存到 HBase 中。

1. 创建表：create

由于将 scores 表的"name"作为行键，所以在创建表时不用预指定行键这一列。并且"时间戳"这一列也是由 HBase 自动生成，所以只需指定两个列族"grade"和"course"。

create 命令的语法格式：

语法：

```
create ' 表名称 ',' 列名称 1',' 列名称 2',…,' 列名称 N'
```

其中表名、列名必须用单引号括起来并以逗号分隔。按照 create 语法操作如下：

```
hbase(main):001:0> create 'scores','grade','course'
0 row(s) in 0.5940 seconds

=> Hbase::Table - scores
```

2. 查看所有表：list

语法：

```
list
```

使用 list 命令可以查看当前 HBase 数据库中所有表，具体操作如下：

```
hbase(main):001:0> list
TABLE
scores
1 row(s) in 0.3580 seconds
```

可以看到当前数据库中已经存在"scores"表。如果要查看该表所有列族的详细描述信息可使用 describe 命令：

语法：

```
describe ' 表名 '
```

例如：

```
hbase(main):002:0> describe 'scores'
Table scores is ENABLED
scores
COLUMN FAMILIES DESCRIPTION
{NAME => 'course', DATA_BLOCK_ENCODING => 'NONE', BLOOMFILTER => 'ROW',
   REPLICAT
ION_SCOPE => '0', VERSIONS => '1', COMPRESSION => 'NONE', MIN_VERSIONS => '0', T
TL => 'FOREVER', KEEP_DELETED_CELLS => 'FALSE', BLOCKSIZE => '65536', IN_MEMORY
=> 'false', BLOCKCACHE => 'true'}
{NAME => 'grade', DATA_BLOCK_ENCODING => 'NONE', BLOOMFILTER => 'ROW',
   REPLICATI
ON_SCOPE => '0', VERSIONS => '1', COMPRESSION => 'NONE', MIN_VERSIONS => '0', TT
L => 'FOREVER', KEEP_DELETED_CELLS => 'FALSE', BLOCKSIZE => '65536', IN_MEMORY =
> 'false', BLOCKCACHE => 'true'}
2 row(s) in 0.1120 seconds
```

其中关于列族描述信息具体含义如表 6-7 所示。

表 6-7 列族描述信息

列族参数	取值	说明
NAME	可打印的字符串	列族名称，参考 ASCII 码表中可打印字符
DATA_BLOCK_ENCODING	NONE（默认）	数据块编码
BLOOMFILTER	NONE（默认）|ROWCOL|ROW	提高随机读的性能
REPLICATION_SCOPE	默认 0	开启复制功能
VERSIONS	数字	列族中单元时间版本最大数量
COMPRESSION	NONE（默认）|LZO|SNAPPY|GZIP	压缩编码
MIN_VERSIONS	数字	列族中单元时间版本最小数量
TTL	默认 FOREVER	单元时间版本超时时间，可为其指定多长时间（秒）后失效
KEEP_DELETED_CELLS	TRUE|FALSE（默认）	启用后避免被标记为删除的单元从 HBase 中移除
BLOCKSIZE	默认 65536 字节	数据块大小。数据块越小，索引越大
IN_MEMORY	true|false，默认 false	使得列族在缓存中拥有更高优先级
BLOCKCACHE	true|false，默认 true	是否将数据放入读缓存

在创建表时，除了列族名称，列族其余的参数均为可选项，上面创建"scores"表的语句为简化的方式，完整的写法为：

create 'scores',{NAME=>'grade',VERSIONS=>5},{NAME=>'course',VERSIONS=5}

对比前面的方式，此命令指定了列族名称及可保存的单元时间版本最大数量。可以看出，指定列族参数的格式为是：参数名 => 参数值，注意赋值符号为"=>"且参数名必须大写，如果指定了多个参数，应用逗号分开，最后所有参数以"{}"括起来表示一个列族。

3. 添加数据：put

向 scores 中增加一些数据，使用 put 命令可向表中插入数据，语法格式为：
语法：

put ' 表名称 ',' 行键 ',' 列键 ',' 值 '

具体操作如下：

```
hbase(main):003:0> put 'scores','tom','grade:','1'
0 row(s) in 0.4090 seconds

hbase(main):004:0> put 'scores','tom','course:art','80'
0 row(s) in 0.0330 seconds

hbase(main):005:0> put 'scores','tom','course:math','89'
0 row(s) in 0.0130 seconds

hbase(main):006:0> put 'scores','jason','grade:','2'
0 row(s) in 0.0220 seconds

hbase(main):007:0> put 'scores','jason','course:art','87'
0 row(s) in 0.0110 seconds

hbase(main):008:0> put 'scores','jason','course:math','57'
0 row(s) in 0.0090 seconds
```

4. 扫描表：scan

scan 用于进行全表单元扫描。
语法：

scan ' 表名称 ',{COLUMNS=>[' 列族名 1',' 列族名 2'…], 参数名 => 参数值…}

大括号内的内容为扫描条件，如不指定则查询所有数据：

```
hbase(main):009:0> scan 'scores'
ROW              COLUMN+CELL
 jason           column=course:art, timestamp=1468390806387, value=87
 jason           column=course:math, timestamp=1468390815858, value=57
```

```
 jason           column=grade:, timestamp=1468390792369, value=2
 tom             column=course:art, timestamp=1468390759906, value=80
 tom             column=course:math, timestamp=1468390774316, value=89
 tom             column=grade:, timestamp=1468390747193, value=1
2 row(s) in 0.1140 seconds
```

输出结果显示共 2 行数据，因为在 scan 的结果中，相同的行键的所有单元视为一行。如果对有些列族不关心，便可指定查询某个列族：

```
hbase(main):010:0> scan 'scores',{COLUMNS=>'course'}
ROW              COLUMN+CELL
 jason           column=course:art, timestamp=1468390806387, value=87
 jason           column=course:math, timestamp=1468390815858, value=57
 tom             column=course:art, timestamp=1468390759906, value=80
 tom             column=course:math, timestamp=1468390774316, value=89
2 row(s) in 0.0260 seconds
```

能不能指定列键来扫描呢？肯定是可以的，语法如下：

scan ' 表名称 ',{COLUMN=>[' 列键 1',' 列键 2'…], 参数名 => 参数值…}

将 COLUMNS 替换成 COLUMN，表示当前扫描的目标是列键，注意区分大小写。如下所示，扫描所有行的列键为"course:math"的单元，并使用 LIMIT 参数限制为输出一个单元。

```
hbase(main):037:0> scan 'scores',{COLUMN=>'course:math',LIMIT=>1}
ROW              COLUMN+CELL
 jason           column=course:math, timestamp=1468392365947, value=57
1 row(s) in 0.0130 seconds
```

5. 获取数据：get

get 用于获取行的所有单元或者某个指定的单元。

语法：

get ' 表名称 ',' 行键 ',{COLUMNS=>[' 列族名 1',' 列族名 2'…], 参数名 => 参数值…}
get ' 表名称 ',' 行键 ',{COLUMN=>[' 列键 1',' 列键 2'…], 参数名 => 参数值…}

与 scan 相比多一个参数即行键。scan 查找的目标是全表的某个列族、列键，而 get 查找的目标是某行的某个列族、列健。

（1）查找行键为"jason"的所有单元：

```
hbase(main):011:0> get 'scores','jason'
COLUMN           CELL
 course:art      timestamp=1468390806387, value=87
 course:math     timestamp=1468390815858, value=57
 grade:          timestamp=1468390792369, value=2
3 row(s) in 0.0620 seconds
```

从输出结果可见，不指定列族或列键，会输出行键的所有列键单元。

（2）精确查找行键为"jason"，列键为"course:math"的单元：

```
hbase(main):012:0> get 'scores','jason',{COLUMN=>'course:math'}
COLUMN              CELL
 course:math           timestamp=1468390815858, value=57
1 row(s) in 0.0140 seconds
```

> **提示**
> - get 'scores','jason',{COLUMNS=>'course'} 等价于 get 'scores','jason','course'
> - get 'scores','jason',{COLUMN=>'course:math'} 等价于 get 'scores','jason','course:math'
> - get 'scores','jason',{COLUMNS=>['course', 'grade']} 等价于 get 'scores','jason','course', 'grade'
> - get 'scores','jason',{COLUMN=>['course:math', 'grade:']} 等价于 get 'scores','jason','course:math', 'grade:'

6. 删除数据：delete

语法：

```
delete ' 表名称 ',' 行键 ',' 列键 '
deleteall ' 表名称 ',' 行键 '
```

delete 只能删除一个单元，而 deleteall 为删除一行。下面删除 scores 表中，行键为"jason"，列键为"course:art"的单元：

```
hbase(main):013:0> delete 'scores','jason','course:art'
0 row(s) in 0.1050 seconds
hbase(main):014:0> get 'scores','jason'
COLUMN              CELL
 course:math           timestamp=1468390815858, value=57
 grade:                timestamp=1468390792369, value=2
2 row(s) in 0.0450 seconds
```

7. 修改表：alter

使用 alter 可为表增加或修改列族。

语法：

```
alter ' 表名称 ', 参数名 => 参数值 ,...
```

其中列族名参数 NAME 必须提供，如果已存在则修改，否则增加一个列族。下面示例将 scores 表的列族"course"的"VERSIONS"参数修改为"5"：

hbase(main):020:0> alter 'scores',NAME=>'course',VERSIONS=>'5'
Updating all regions with the new schema...
0/1 regions updated.
1/1 regions updated.
Done.
0 row(s) in 3.6470 seconds

同时修改或增加多个列族时以逗号分开，并且每个列族用"{}"括起来。
语法：

alter ' 表名称 ',{ 参数名 => 参数值 ,...},{ 参数名 => 参数值 ,...}…

下面示例将同时修改 scores 表的两个列族：

hbase(main):021:0> alter 'scores',{NAME=>'grade',VERSIONS=>'5'},{NAME=>'course', VERSIONS=>'5'}
Updating all regions with the new schema...
0/1 regions updated.
1/1 regions updated.
Done.
Updating all regions with the new schema...
0/1 regions updated.
1/1 regions updated.
Done.
0 row(s) in 4.4030 seconds

8. 删除表：drop

在前面的 describe 命令操作过程中就可以发现，HBase 表分两种状态：DISABLED 和 ENABLED，分别表示是否为可用状态。

使用 disable 将表置为不可用状态：

hbase(main):022:0> disable 'scores'
0 row(s) in 1.4680 seconds

使用 disable 将表置为可用状态：

hbase(main):023:0> enable 'scores'
0 row(s) in 0.5770 seconds

当表为 ENABLED 状态时，表会禁止被删除。所以必须先将表置为 DISABLED 状态。操作如下：

hbase(main):024:0> disable 'scores'
0 row(s) in 1.2650 seconds

hbase(main):025:0> drop 'scores'
0 row(s) in 0.5030 seconds

hbase(main):026:0> list

TABLE
0 row(s) in 0.0200 seconds

通过以上 8 个命令，我们就可对 HBase 数据库表进行基本管理，对表数据进行增加、删除、查询操作。而对于表单元的修改实质上是增加操作，HBase 保留了单元的多个版本，默认查询最新版本。

6.3 MapReduce 与 HBase

为什么要集成 MapReduce 和 HBase？

我们知道，HBase 可以使用本地文件系统和 HDFS 文件系统作为数据存储介质，当在伪分布式和完全分布式下运行时，其使用的是 HDFS 文件系统。我们不用关心 HBase 中的表是如何在 HDFS 上存储的，但我们知道的是数据最终会被写入某些文件中，并且我们可以通过 HBase 将数据从这些文件中读取出来。

再来看，一个 MapReduce 应用要被定义为一个作业才能在 MapReduce 框架中运行，这些定义包括两个基本要素：MapReduce 的输入和输出，包括数据输入/输出的文件和处理这些文件所采用的输入/输出格式。

综上两点，可以让 MapReduce 作业需要输入数据时从 HBase 中读取，而在输出数据时，又可以输出到 HBase 完成存储，达到 HBase 与 MapReduce 协同工作，为 MapReduce 提供数据输入输出的目的。这样带来的好处是，我们既利用了 MapReduce 的分布式计算的优势，也利用了 HDFS 海量存储的特点，特别是利用了 HBase 对海量数据的实时访问的特点。通过 MapReduce 和 HBase 的集成，MapReduce、HBase、HDFS 之间关系如图 6.14 所示。

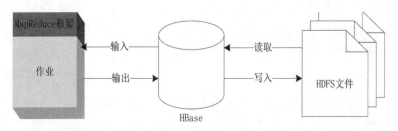

图 6.14 MapReduce 与 HBase 集成

除了将 HBase 作为 MapReduce 作业的输入和输出，集成 MapReduce 与 HBase 还可以做什么呢？

（1）可以对 HBase 中的数据进行非实时性的统计分析。HBase 适合做 Key-Value 查询，默认不带聚合函数（sum、avg 等），对于这种需求非常适合集成 MapReduce 来完成，但我们也应该注意到 MapReduce 的局限性，MapReduce 的本身高延迟使得它不能满足实时交互式的计算。

（2）可以对 HBase 的表数据进行分布式计算。HBase 的目标是在海量数据中快速定位所需要的数据并访问它，可以发现 HBase 只能按照行键查询并不支持其他条件查询，所以我们只依靠 HBase 来解决存储的扩展，而不是业务逻辑，那么此时将业务逻辑放到 MapReduce 计算框架中是合适的。

（3）可以在多个 MapReduce 间使用 HBase 作为中间存储介质。

具体怎么进行集成呢？

HBase Java API 对 MapReduce API 进行了扩展，这里将其称为 HBase MapReduce API。显然这是由 HBase 提供，主要是为了方便 MapReduce 应用对 HTable 的操作。

前面提过，MapReduce 的输入和输出包括数据输入/输出的文件和处理这些文件所采用的输入/输出格式，与 HBase 集成后，"输入/输出的文件"变为"表（HTable）"，那么针对表的输入/输出格式也得提供相应实现，分别是 TableInputFormat 和 TableOutputFormat，其所在 jar 文件为"hbase-server-1.0.2.jar"。同时，HBase 还提供了 TableMapper 和 TableReducer 类使得编写 MapReduce 程序更加方便。

表 6-8 是 HBase MapReduce API 主要类与 Hadoop MapReduce API 的对应关系。

表 6-8　HBase MapReduce API 与 MapReduce API

HBase MapReduce API	Hadoop MapReduce API
org.apache.hadoop.hbase.mapreduce.TableMapper	org.apache.hadoop.mapreduce.Mapper
org.apache.hadoop.hbase.mapreduce.TableReducer	org.apache.hadoop.mapreduce.Reducer
org.apache.hadoop.hbase.mapreduce.TableInputFormat	org.apache.hadoop.mapreduce.InputFormat
org.apache.hadoop.hbase.mapreduce.TableOutputFormat	org.apache.hadoop.mapreduce.OutputFormat

表 6-8 中左侧的类均继承于右侧的类，关于 HBase MapReduce API 的具体使用在稍后再详细讲解。默认情况下，MapReduce 作业发布到集群中后，不能访问 HBase 的配置文件和相关类，所以首先需要对集群中的各节点的 Hadoop 环境做如下调整：

（1）将 hbase-site.xml 复制到 $HADOOP_HOME/etc/hadoop 下。

（2）编辑 $HADOOP_HOME/etc/hadoop/hadoop-env.sh，增加一行：

export HADOOP_CLASSPATH=$HADOOP_CLASSPATH:~/hbase-1.0.2/lib/*

其中，HBase 的路径视其安装目录而不同。第一步可使 MapReduce 作业在运行时连接到 ZooKeeper 集群；第二步将 HBase 安装目录 lib 下的所有 jar 文件添加到环境变量 $HADOOP_CLASSPATH 中，使得 MapReduce 作业可以访问所依赖的 HBase 相关类，从而不用每次将 HBase 相关类打包到 MapReduce 应用的 jar 文件中。最后注意将上述操作的两个文件，复制到 Hadoop 集群中的其他节点上。使用如下命令可测试环境是否已正确配置：

hadoop jar ~/hbase-1.0.2/lib/hbase-server-1.0.2.jar rowcounter music

该命令将运行"hbase-server-1.0.2.jar"中的 MapReduce 应用"rowcounter"，参数

为表名"music"。其功能是使用 MapReduce 框架统计 HBase 数据库表 music 中的行数。运行结束后输出如图 6.15 所示表示环境已正确配置。

```
org.apache.hadoop.hbase.mapreduce.RowCounter$RowCounterMapper$Counters
        ROWS=2
File Input Format Counters
        Bytes Read=0
File Output Format Counters
        Bytes Written=0
```

图 6.15　测试 MapReduce 和 HBase 集成环境

6.4　Hive 和 Spark

完整的大数据平台应该提供离线计算、实时计算、实时查询这几方面的功能。离线计算就是非实时计算，通常这类计算要在开始前就知道问题的所有数据输入。MapReduce 就是典型的离线计算，用于对全部归档数据进行批量处理，然后将结果缓存起来提供查询（如 HBase），可以看出从数据输入到结果输出的整个阶段，MapReduce 在实时性上并不是非常理想。

在实际运用中通常采用 Hadoop+Spark+Hive（MapReduce）的解决方案。利用 Hadoop 的 HDFS 解决分布式存储问题；利用 MapReduce 或 Hive 解决离线计算问题；利用 Spark 解决实时计算问题；最后利用 HBase 来解决实时查询的问题。

6.4.1　Hive

Hive 是 Hadoop 中的一个重要子项目，它的优势在于可以利用 MapReduce 编程技术，提供了类似 SQL 的编程接口，实现部分 SQL（结构化查询语句）语句的功能。Hive 的出现极大地推进了 Hadoop 在数据仓库方面的发展。

Hive 定义了类 SQL 的语言——HiveQL。使用 HiveQL 意味着，不需要编写 MapReduce 就可以方便地使用 Mapper 和 Reducer 操作，这对 MapReduce 框架是一个强有力的支持。

Hive 本身建立在 Hadoop 体系结构上，提供了一个 SQL 解析的过程，从外部接口中获取命令，并对用户指令进行解析。Hive 将外部命令解析成一个 MapReduce 作业，随后提交到 Hadoop 集群进行处理。

Hive 的出现，是要解决如何让用户从一个现有的数据基础架构转移到 Hadoop 上，而这个基础架构是基于传统关系型数据库和 SQL 的。大多数的数据仓库应用程序是使用基于 SQL 的关系型数据库实现的，所以 Hive 降低了将这些应用移植到 Hadoop 上的障碍。用户如果懂得 SQL，那么学习使用 Hive 将会很容易，否则只能重新学习 MapReduce 编程。

典型 Hive 命令如下：

```
hive>CREATE TABLE employees(name STRING,salary FLOAT);
hive>SELECT name ,salary FROM employees;
```

Hive 的默认文件格式为以行存储的文本格式，文件中每行表示一个记录，记录之间以不同分隔符来区别。如下所示：

```
Jason^A8000
Tom^A7000
...
```

其中"^A"表示字段间分隔符。

更多详细资料请参考 http://hive.apache.org。

6.4.2 Spark

Apache Spark 是一个新出现的大数据处理引擎，和 Hadoop 都属于大数据解决方案，相同之处是 Spark 也提供了类似 MapReduce 的处理，但是 Spark 没有提供文件管理系统，所以它必须和其他的分布式文件系统进行集成才能运行，实际使用中，通常是选择 Hadoop 的 HDFS，并让 Spark 运行在 YARN 上。

首先看下 Hadoop MapReduce 的局限和不足：

- 只提供两个操作：Mapper 和 Reducer，表达力欠缺。
- 复杂的计算需要大量的 Job，Job 间的依赖关系由开发者自己管理。
- 中间结果也放在 HDFS 中。
- ReduceTask 需要等待所有 MapTask 都完成后才可以开始。
- 延时高，不能实时计算。
- 对于迭代式计算性能差。

MapReduce 批量处理操作顺序为：

（1）从集群读取数据。
（2）然后对数据进行操作。
（3）将结果写回到集群。
（4）从集群读取更新后的数据，执行下一个数据操作。

Spark 执行类似的操作，不过是在内存中执行。所以 Spark 对内存的要求比 MapReduce 要大得多。总的来说，Spark 的优势在于处理数据的速度比 MapReduce 要快很多，API 简单易用，尤其在交互式查询、流式计算及迭代式计算方面性能有很大提升。

实际上，Spark 与 Hadoop 是一种相互共生的关系。Hadoop 提供了 Spark 所没有的功能特性，比如分布式文件系统，而 Spark 为需要它的那些数据集提供了实时内存处理。

更多详细资料请参考 http://spark.apache.org。

本章总结

- HBase 能够实时地随机访问超大规模数据集，是对 HDFS 有力的补充。
- HBase 体系结构中的主体包括：HRegion、HRegionServer、HMaster、ZooKeeper。
- HBase 数据模型包括：表、行（行键）、列族、列（列键）、单元格、时间戳。
- HBase 概念视图看每个表由很多行组成，但在实际存储时，是按照列族来存储的。
- HBase 可以在单机、伪分布式和完全分布式三种模式下运行。
- HBase Shell 常用命令有：alter、count、create、describe、delete、deleteall、disable、drop、enable、exists、exit、get、incr、list。

本章作业

1. 关系型数据库与 HBase 数据库有哪些区别？
2. 在 HBase 的 scores 表中，此前 jason 的 course:math 成绩为 57 分，现在补考成绩为 63，请使用 put 命令增加补考成绩，然后使用 scan 命令显示 jason 的 course:math 成绩的所有版本。

   提示
   - alter 'scores',NAME=>'course',VERSIONS=>5 可修改最大版本数量。
   - scan ' 表名 ',{COLUMN=>' 列键 ',VERSIONS=>5} 最多可显示最近的 5 个版本。
   - scan 的参数 STARTROW 和 ENDROW 可以查询行的范围。
3. 用课工场 APP 扫一扫完成在线测试，快来挑战吧！

第7章

部署 CDH 环境

技能目标

- 理解 CDH 核心概念
- 会进行 CDH 集群的部署
- 会使用管理控制台对 CDH 集群进行管理
- 会使用添加 CDH 集群服务

本章导读

  Cloudera Hadoop（CDH）是 Cloudera 公司的发行版本，基于 Apache Hadoop 的二次开发，优化了组件兼容和交互接口、简化安装配置、增加 Cloudera 兼容特性。CDH 常用安装方式包括：Cloudera Manager 在线安装、Parcel 安装、YUM 安装以及 RPM 安装。官方文档推荐使用 Parcel 模式进行安装，由于官方版本众多，要注意根据系统选择对应版本，避免安装过程中出现错误。

知识服务

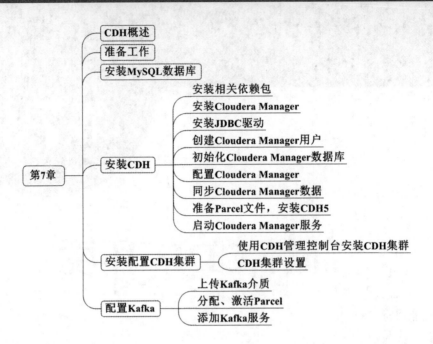

7.1 CDH 概述

Apache Hadoop 是目前最主流的在通用硬件构建大型集群上运行应用程序的分布式架构。采用 Apache 2.0 许可协议发布开源协议，从而用户可以免费使用以及任意修改 Hadoop。官方版本称为社区版 Hadoop，市面上有很多其他 Hadoop 版本，比较流行的有 2 个版本，Apache 版本和 Cloudera 版本。

- Apache Hadoop：维护人员比较多，更新频率比较快，稳定性相对比较差。
- Cloudera Hadoop（CDH）：是 Cloudera 公司的发行版本，基于 Apache Hadoop 的二次开发，优化了组件兼容和交互接口、简化安装配置、增加 Cloudera 兼容特性。

7.2 案例环境

本案例使用 CentOS 7.2 操作系统，具体如表 7-1 所示。

表 7-1 案例环境

主机名	IP 地址	内存
node1.master	192.168.0.102	至少 6G
node2.slave	192.168.0.104	至少 4G
node3.slave	192.168.0.105	至少 4G

CDH 常用安装方式包括：Cloudera Manager 在线安装、Parcel 安装、YUM 安装以及 RPM 安装。官方文档推荐使用 Parcel 模式进行安装，故本次安装采用 Percel 方式进行。这里先下载离线资源，由于官方版本众多，一定要注意根据系统选择对应版本，若版本选择有误安装过程中也会出现错误。

以下是本案例使用的安装介质：

（1）Cloudera Manager

可以从 http://archive.cloudera.com/cm5/cm/5/ 地址进行下载，这里下载的版本是 cloudera-manager-centos7-cm5.9.9_x86_64.tar.gz。

（2）JDBC 驱动

可以从 MySQL 官方网站 http://dev.mysql.com/downloads/connector/j/ 上下载最新的 MySQL JDBC 驱动程序，这里下载的是 mysql-connector-java-5.1.40.tar.gz。

（3）CDH Percel 包

可以从 http://archive.cloudera.com/cdh5/parcels/ 地址进行下载，分别需要下载三个文件，这里下载的是 CDH-5.9.0-1.cdh5.9.0.p0.23-el7.parcel、CDH-5.9.0-1.cdh5.9.0.p0.23-el7.parcel.sha1、manifest.json。

（4）Kafka CSD 包

可以从 http://archive.cloudera.com/kafka/parcels/latest/ 地址进行下载，分别需要下载两个文件，这里下载的是 KAFKA-2.0.2-1.2.0.2.p0.5-el7.parcel、KAFKA-2.0.2-1.2.0.2.p0.5-el7.parcel.sha1。

7.2.1 准备工作

1. 修改主机名

分别配置各个节点的主机名。

IP 地址为 192.168.0.102 的主机：

[root@localhost ~]# hostnamectl set-hostname node1.master

IP 地址为 192.168.0.104 的主机：

 [root@localhost ~]# hostnamectl set-hostname node2.slave

IP 地址为 192.168.0.105 的主机：

[root@localhost ~]# hostnamectl set-hostname node3.slave

2. 关闭防火墙和 SELinux

需要在所有的节点上执行，因为涉及到的端口比较多，临时关闭防火墙是为了安装起来更方便，安装完毕后可以根据需要设置防火墙策略，保证集群安全。

[root@localhost ~]# systemctl stop firewalld
[root@localhost ~]# systemctl disable firewalld

```
[root@localhost ~]# vi /etc/sysconfig/selinux
SELINUX=permissive
[root@localhost ~]# reboot
```

3. 增加 hosts 解析

修改 /etc/hosts 文件，配置主机名与 IP 地址的解析。

```
[root@node1 ~]# vi /etc/hosts
192.168.0.102node1.master
192.168.0.104node2.slave
192.168.0.105 node3.slave
[root@node1 ~]# scp /etc/hosts node2.slave:/etc/
[root@node1 ~]# scp /etc/hosts node3.slave:/etc/
```

4. 同步各节点时间

分别在 node1.master、node2.slave、node3.slave 节点上配置 ntp 服务与网络上的时钟服务器 pool.ntp.org 进行时间同步，以在 node1.msater 上配置为例：

```
[root@node1 ~]# yum install ntp
[root@node1 ~]# vi /etc/ntp.conf
```

添加需要的时钟服务器：

```
server pool.ntp.org iburst
[root@node1 ~]# systemctl start ntpd
[root@node1 ~]# systemctl enable ntpd
[root@node1 ~]# ntpstat
synchronised to NTP server (108.59.2.24) at stratum 3
   time correct to within 8319 ms
   polling server every 64 s
```

5. 安装 JDK 环境

后续要安装 Presto，要求采用 JDK8 以上版本。需要在所有节点上安装 JDK 环境。如果系统中有自带的 JDK，需要先删除自带的 JDK 相关软件包。

```
[root@node1 ~]# tar zxvf jdk-8u91-linux-x64.tar.gz -C /opt
[root@node1 ~]# ln -s /opt/jdk1.8.0_91/ /opt/jdk
```

给 /opt/jdk 创建软链接 /usr/java/default，否则 Spark 将无法安装完成。

```
[root@node1 ~]# mkdir /usr/java
[root@node1 ~]# ln -s /opt/jdk /usr/java/default
```

修改环境变量。

```
[root@node1 ~]# vi /etc/profile
export JAVA_HOME=/opt/jdk
export PATH=$JAVA_HOME/bin:$PATH
[root@node1 ~]# source /etc/profile
```

验证安装的 JDK 版本。

```
[root@node1 ~]# java -version
java version "1.8.0_91"
Java(TM) SE Runtime Environment (build 1.8.0_91-b14)
Java HotSpot(TM) 64-Bit Server VM (build 25.91-b14, mixed mode)
```

6. 设置主节点到其他节点的免密钥登录

在主节点 node1.master 上生成 SSH 密钥对，分别复制公钥到从节点 node2.slave 和 node3.slave 上，实现主节点的免密钥登录从节点。

```
[root@node1 ~]# ssh-keygen -t rsa
Generating public/private rsa key pair.
Enter file in which to save the key (/root/.ssh/id_rsa):
Enter passphrase (empty for no passphrase):
Enter same passphrase again:
Your identification has been saved in /root/.ssh/id_rsa.
Your public key has been saved in /root/.ssh/id_rsa.pub.
The key fingerprint is:
86:18:aa:6b:e9:38:53:4c:5d:a5:64:5a:2e:0c:ee:09 root@node1
The key's randomart image is:
+--[ RSA 2048]----+
|  .  +..         |
| . o *..         |
| E ..=.o         |
| o.o.+ .         |
| o+ . . S        |
| .o .            |
|...              |
|++               |
|=+               |
+-----------------+
[root@node1 ~]# ssh-copy-id node2.slave
[root@node1 ~]# ssh-copy-id node3.slave
```

测试登录。

```
[root@node1 ~]# ssh node2.slave
Last login: Thu Jan 12 14:23:39 2017 from node1.master
[root@node2 ~]# exit
logout
Connection to node2.slave closed.
[root@node1 ~]# ssh node3.slave
Last login: Thu Jan 12 14:24:37 2017 from node1.master
[root@node3 ~]# exit
logout
Connection to node3.slave closed.
```

7. 交换分区和大页设置

分别在三个节点上禁用交换分区和透明大页。否则会在安装配置 CDH 集群环境检测中报错。

```
[root@node1 ~]# sysctl -w vm.swappiness=0
vm.swappiness = 0
[root@node1 ~]# echo never > /sys/kernel/mm/transparent_hugepage/defrag
[root@node1 ~]# echo never>/sys/kernel/mm/transparent_hugepage/enabled
[root@node1~]# echo "echo never > /sys/kernel/mm/transparent_hugepage/defrag" >> /etc/rc.d/rc.local
[root@node1 ~]# echo "echo never>/sys/kernel/mm/transparent_hughugepage/enabled" >> /etc/rc.d/rc.local
[root@node1 ~]# chmod +x /etc/rc.d/rc.local
```

7.2.2 安装数据库

在主节点 node1.master 上安装 MySQL 数据库。CentOS 7 上自带的数据库为 MariaDB，需要从新下载 MySQL 官方社区版本。

```
[root@node1 ~]# rpm -qa |grep mariadb
mariadb-libs-5.5.44-2.el7.centos.x86_64
[root@node1 ~]# rpm -e --nodeps mariadb-libs-5.5.44-2.el7.centos.x86_64
```

从 MySQL 官网 http://dev.mysql.com/downloads/mysql/ 下载 MySQL 软件包进行配置。

```
[root@node1 ~]#tar xvf mysql-5.7.17-1.el7.i686.rpm-bundle.tar
[root@node1 ~]#yum localinstall mysql*
[root@node1 ~]# systemctl start mysqld
[root@node1 ~]# systemctl enable mysqld
```

安装完毕后，会自动生成 MySQL 用户密码，需要修改 MySQL 初始密码。

```
[root@node1 ~]# grep "password" /var/log/mysqld.log
2017-02-04T02:52:53.077397Z 1 [Note] A temporary password is generated for root@localhost: )
    rX5?VuVlJ+o// 初始 root 密码
[root@node1 ~]# mysqladmin -u root -p password
Enter password:
New password:
Confirm new password:
```

输入默认密码，然后输入两次新密码 H@o123.com，因为 MySQL 启用了密码增加插件，新密码必须符合密码复杂性要求。

之后为 Hive、Oozie 创建数据库，并授予 cdh 用户权限。

```
[root@node1 ~]# mysql -u root -p
mysql> CREATE DATABASE hive DEFAULT CHARSET utf8 COLLATE utf8_general_ci;
Query OK, 1 row affected (0.00 sec)
mysql> CREATE DATABASE oozie DEFAULT CHARSET utf8 COLLATE utf8_general_ci;
Query OK, 1 row affected (0.00 sec)
```

```
mysql> grant all privileges on *.* to 'cdh'@'localhost' identified by '123@H.com' with grant option;
Query OK, 0 rows affected (0.00 sec)
mysql> grant all privileges on *.* to 'cdh'@'%' identified by '123@H.com' with grant option;
Query OK, 0 rows affected (0.00 sec)
mysql> flush privileges;
Query OK, 0 rows affected (0.01 sec)
mysql> quit
Bye
```

7.2.3 安装 CDH

1. 安装相关依赖包

在实际安装过程中，CDH 对部分包存在依赖（根据环境差异，具体安装过程中可能存在一定的区别，注意分析查看日志文件）。本案例环境中有以下几个软件包没有安装导致报错，直接使用 YUM 进行安装。

```
[root@node1 ~]# yum install -y psmisc libxslt libxslt-python perl
```

在 node2.slave 与 node3.slave 上也需要安装。

2. 安装 Cloudera Manager

将下载的 cloudera-manager-centos7-cm5.9.9_x86_64.tar.gz 上传到主服务器（node1.master）上，进行解压。

```
[root@node1~]# tar xzvf cloudera-manager-centos7-cm5.9.0_x86_64.tar.gz -C /opt/ [root@node1 ~]
  # mv /opt/cm-5.9.0/ /opt/cm
```

3. 安装 JDBC 驱动

同时将 mysql-connector-java-5.1.40.tar.gz 上传到主服务器，复制到 Cloudera Manager 安装目录下的 cm/share/cmf/lib/ 目录中，并修改权限。

```
[root@node1 ~]# tar xzvf mysql-connector-java-5.1.40.tar.gz
[root@node1 ~]# cd mysql-connector-java-5.1.40
[root@node1 mysql-connector-java-5.1.40]# chmod +x mysql-connector-java-5.1.40-bin.jar
[root@node1 mysql-connector-java-5.1.40]# cp mysql-connector-java-5.1.40-bin.jar /opt/cm/share/
  cmf/lib/mysql-connector-java.jar
[root@node1 mysql-connector-java-5.1.40]# cp mysql-connector-java-5.1.40-bin.jar /usr/share/java/
  mysql-connector-java.jar
```

4. 创建 Cloudera Manager 用户

注意是在所有节点上均需要创建 Cloudera Manager 用户。

```
[root@node1 ~]# useradd --system --home=/opt/cm/run/cloudera-scm-server/ --no-create-home
  --shell=/bin/false --comment "Cloudera SCM User" cloudera-scm
```

5. 初始化 Cloudera Manager 数据库

使用 Cloudera Manager 的 scm_prepare_database.sh 初始化数据库，数据库名为 cm，对应的用户名、密码为 scm 和 123456。

```
[root@node1 ~]# /opt/cm/share/cmf/schema/scm_prepare_database.sh mysql cm -hlocalhost -ucdh
    -p'123@H.com' scm '123@H.com'
JAVA_HOME=/opt/jdk
Verifying that we can write to /opt/cm/etc/cloudera-scm-server
Creating SCM configuration file in /opt/cm/etc/cloudera-scm-server
Executing: /opt/jdk/bin/java -cp /usr/share/java/mysql-connector-java.jar:/usr/share/java/oracle-
    connector-java.jar:/opt/cm/share/cmf/schema/../lib/* com.cloudera.enterprise.dbutil.
    DbCommandExecutor /opt/cm/etc/cloudera-scm-server/db.properties com.cloudera.cmf.db.
[main] DbCommandExecutor INFO Successfully connected to database.
All done, your SCM database is configured correctly!
```

6. 配置 Cloudera Manager

修改 Cloudera Manager 的配置文件，将 server_host 改为主节点 node1.master 主机。

```
[root@node1 ~]# vi /opt/cm/etc/cloudera-scm-agent/config.ini
server_host=192.168.0.102
```

7. 同步 Cloudera Manager 数据到其他节点

```
[root@node1 ~]# scp -r /opt/cm/ node2.slave:/opt/
[root@node1 ~]# scp -r /opt/cm/ node3.slave:/opt/
```

这里需要注意，如果在同步到其他节点之前，启动过代理程序，需要删除所有服务器上的 /opt/cm/lib/cloudera-scm-agent 中生成的 response.avro 和 uuid 两个文件，并重启代理程序，否则服务程序将无法正确检查到 agent。

8. 准备 Parcel 文件，安装 CDH5

上传 Parcel 文件到主节点服务器的 /opt/cloudera/parcel-repo/ 目录中，同时将 CDH-5.9.0-1.cdh5.9.0.p0.23-el7.parcel.sha1 重命名为 CDH-5.9.0-1.cdh5.9.0.p0.23-el7.parcel.sha，否则系统会重新下载 CDH-5.9.0-1.cdh5.9.0.p0.23-el7.parcel.sha 文件。

```
[root@node1 ~]# mv manifest.json CDH-5.9.0-1.cdh5.9.0.p0.23-el7.parcel CDH-5.9.0-1.
    cdh5.9.0.p0.23-el7.parcel.sha1 /opt/cloudera/parcel-repo/
[root@node1 ~]# cd /opt/cloudera/parcel-repo/
[root@node1 parcel-repo]#mv CDH-5.9.0-1.cdh5.9.0.p0.23-el7.parcel.sha1 CDH-5.9.0-1.
    cdh5.9.0.p0.23-el7.parcel.sha
```

9. 启动 Cloudera Manager 服务

在 node1.master 节点启动 Cloudera Manager 服务和代理服务。

```
[root@node1 ~]# /opt/cm/etc/init.d/cloudera-scm-server start
Starting cloudera-scm-server:                    [ OK ]
```

```
[root@node1 ~]# /opt/cm/etc/init.d/cloudera-scm-agent start
Starting cloudera-scm-agent:                    [  OK  ]
```

以上脚本是启动命令，停止或重启命令仅需将 start 变成 stop 或 restart 即可。

在 node2.slave 和 node3.slave 节点上启动 Cloudera Manager 代理服务。

```
[root@node2 ~]# /opt/cm/etc/init.d/cloudera-scm-agent start
Starting cloudera-scm-agent:                    [  OK  ]
[root@node3 ~]# /opt/cm/etc/init.d/cloudera-scm-agent start
Starting cloudera-scm-agent:                    [  OK  ]
```

7.2.4　安装配置 CDH 集群

Cloudera Manager Server 和 Agent 服务都启动后，就可以进行 CDH5 的安装配置了。

1．使用 CDH 管理控制台

使用浏览器打开 http://192.168.0.102:7180 登录 Cloudera Manager 的 Web 管理控制台，如图 7.1 所示（默认的用户名和密码为 admin/admin）。

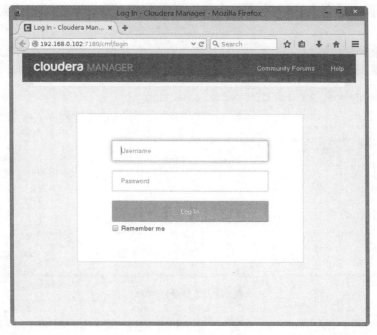

图 7.1　Web 管理控制台

第一次进入时，需选择接受用户授权协议，之后需要选择安装 CDH 的版本，如图 7.2 所示。这里选择 Cloudera Express，单击"继续"按钮，进行 CDH 集群配置。

第一次配置 CDH 集群会自动启动安装向导进行配置安装。

在"Currently Managed Hosts"标签即当前管理的主机，勾选所有的服务器，成为安装集群主机节点，单击"继续"按钮，如图 7.3 所示。

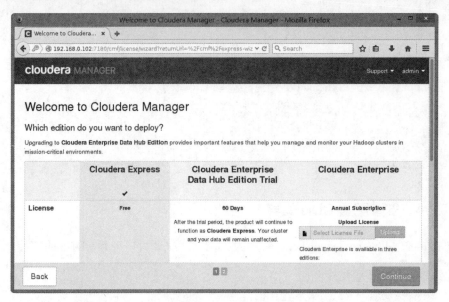

图 7.2 CDH 版本

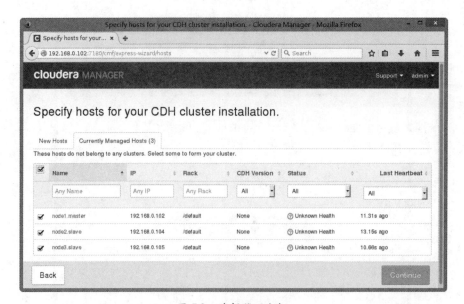

图 7.3 选择管理主机

注意只有已启动 cloudera-scm-agent 的主机才能被 Cloudera Manager 管理。

也可以在"New Hosts"标签，指定主机进行配置的主机，有多种指定方式：直接列出 IP 地址或主机名，多台主机可以以逗号、分号、制表符、空格或放置在单独的行。指定范围例如：10.0.222.[5-7] 或 node1.master,node[2-3].slave，并且需要关闭指定主机防火墙。

选择需要安装的 Parcel 包，单击"继续"按钮进入下一步，如图 7.4 所示。若没有，检查相关文件是否被正确放入安装目录的 cloudera/parcel-repo/ 中。

图 7.4　选择 Parcel 包

之后会将 Parcel 分发到三台服务器，并进行解压、激活操作，该步骤会花费一定的时间，如图 7.5 所示。等待操作完成后，单击"继续"按钮即可。

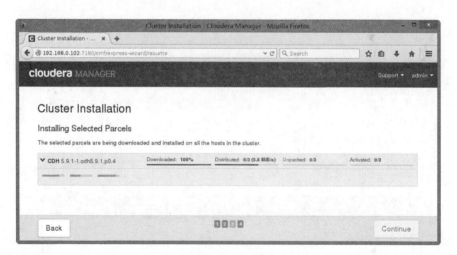

图 7.5　Parcel 分发

在激活完成后，HIVE 以及 Oozie 安装需要使用到 MySQL JDBC 驱动，需要在所有节点上执行以下操作。

```
[root@node1 ~]# cd mysql-connector-java-5.1.40
[root@node1 mysql-connector-java-5.1.40]# cp mysql-connector-java-5.1.40-bin.jar /opt/cloudera/
    parcels/CDH/lib/hive/lib/mysql-connector-java.jar
[root@node1 mysql-connector-java-5.1.40]# cp mysql-connector-java-5.1.40-bin.jar /opt/cloudera/
    parcels/CDH/lib/hadoop/mysql-connector-java.jar
[root@node1 mysql-connector-java-5.1.40]# cp mysql-connector-java-5.1.40-bin.jar  /var/lib/oozie/
    mysql-connector-java.jar
```

在 node2.slave 与 node3.slave 上执行相同指令。

之后检查到安装环境是否满足要求,如安装过程依据环境准备完成了所有的准备工作,将无提示信息,如图 7.6 所示。如果有警告提示内容,按照提示进行修正即可,完成后进行重新检测,全部修正后,单击"完成"按钮进入正式安装过程。

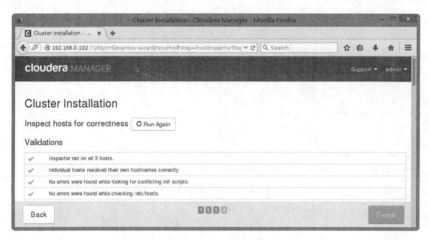

图 7.6　检查安装环境

2. 集群设置

以下是正式的 CDH 集群设置。需要通过以下几个步骤。

首先选择安装 CDH 集群的服务,这里仅安装需要的 HDFS、YARN、Oozie、Hive 等组件,如图 7.7 所示。安装太多不需要的服务在运行过程中会占用较多资源,且选择服务过多到下一步初始化和启动各节点角色的时候会花费大量时间。

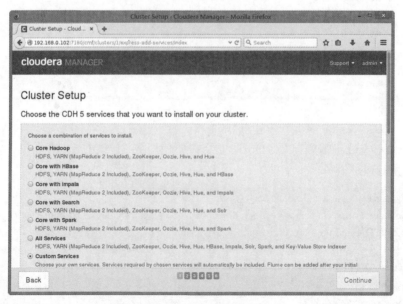

图 7.7　选择安装服务

这里不必担心没安装的服务以后有需要怎么办，在 Cloudera Manager 管理控制台中还可以随时对各节点的服务进行添加和变更。

然后将不同服务分配到不同服务器上，如图 7.8 所示，单击"继续"按钮进入下一步。

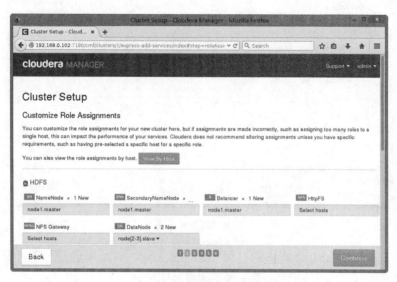

图 7.8　选择安装服务器节点

之后，需要输入 Hive、Oozie 的数据源配置，按照之前在 MySQL 数据库中所创建的数据填写，如图 7.9 所示，单击"测试连接"按钮，测试通过后，单击"继续"按钮。

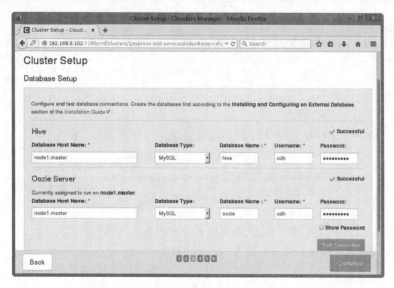

图 7.9　填写数据库信息

之后，依据要求，调整系统配置参数，如图 7.10 所示，调整后单击"继续"按钮。这里也使用默认参数。

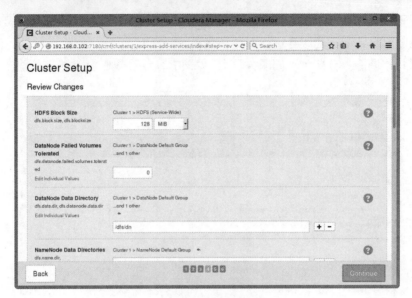

图 7.10　调整配置参数

之后 Cloudera Manager 将依据前面做的配置内容，对环境进行部署，服务安装完成后，所有服务会自动启动，按照向导完成安装即可。

如本步骤出现问题，注意到 Cloudera Manager 的运行日志中，查找对应的日志信息，具体目录为 /opt/cm/run/cloudera-scm-agent/process。成功安装之后，进入 Cluster 页面，如图 7.11 所示。

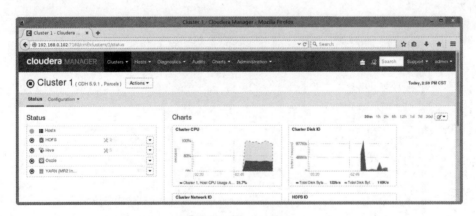

图 7.11　Cluster 页面

7.2.5　配置 Kafka

1. 上传 Kafka 介质

将 KAFKA-1.2.0.jar 上传到服务器的 /opt/cloudera/csd 目录中，将 KAFKA-2.0.0-1.kafka2.0.0.p0.12-el7.parcel、KAFKA-2.0.0-1.kafka2.0.0.p0.12-el7.parcel.sha1 上传到服务器

的 /opt/cloudera/parcel-repo 目录中。将 KAFKA-2.0.0-1.kafka2.0.0.p0.12-el7.parcel.sha1 复制为 KAFKA-2.0.0-1.kafka2.0.0.p0.12-el7.parcel.sha。

```
[root@node1 ~]# mvKAFKA-1.2.0.jar/opt/cloudera/csd
[root@node1 ~]# mvKAFKA-2.0.2-1.2.0.2.p0.5-el7.parcel/opt/cloudera/parcel-repo
[root@node1 ~]#mv KAFKA-2.0.2-1.2.0.2.p0.5-el7.parcel.sha1 /opt/cloudera/parcel-repo/KAFKA-2.0.2-1.2.0.2.p0.5-el7.parcel.sha
```

2. 分配、激活 Parcel

在 Cloudera Manager 管理控制台"Hosts"选项卡中的"Parcel"界面，检查新的 Parcel 后，分配并激活 KAFKA 对应的 Parcel，如图 7.12 所示。

图 7.12　激活 KAFKA 的 Parcel

3. 添加 Kafka 服务

在"Cluster"选项卡中，从"Actions"下拉列表中选择"Add Service"进入添加服务页面，选择需要增加 Kafka 服务，如图 7.13 所示，如果之前没有安装 ZooKeeper 服务，需要同时勾选。

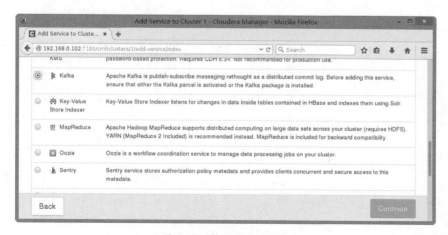

图 7.13　增加 Kafka 服务

之后依据向导完成后续确认操作即可。

后续新增服务可以继续在"Cluster"选项卡,从"Actions"下拉列表中选择"Add Service"进行添加即可。

本章总结

- 在 Hadoop 的各类版本中,比较流行的有 Apache Hadoop 和 Cloudera Hadoop (CDH)。
- CDH 常用安装方式有 Cloudera Manager 在线安装、Parcel 安装、YUM 安装以及 RPM 安装。
- 可以使用 Cloudera Manager 的 Web 管理控制台对 CDH 集群进行配置与管理。

第8章

云计算数据中心与亚马逊AWS

技能目标

- 了解数据中心的网络应用架构
- 了解数据中心的容灾系统
- 了解亚马逊AWS的服务及使用

本章导读

云计算是继大型计算机到客户端—服务器模式转变之后的又一次重大变革,能够快速调动系统中所有可利用资源以提供高性能运算服务,彻底改变了传统计算与存储方式。

云计算数据中心是将多个数据中心进行整合,形成一个统一的云计算数据中心,其中云计算服务是云计算数据中心的外在实现,内部由云计算平台进行支撑,通过各种虚拟化技术构建一个整体的、标准化、自动化基础设施环境和高可用计算环境,综合云计算和数据中心二者的优势为数据调配提供所需的强大计算能力,以提高企业信息化服务效率。目前国内外已经有众多采用云计算技术建立的数据中心。

知识服务

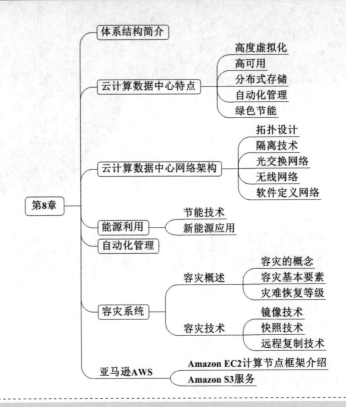

8.1 体系结构简介

云计算数据中心总体架构分为服务和管理两部分。在服务部分为云用户提供基于云平台的各种环境，主要包含三个层次内容：IaaS（基础设施即服务）、PaaS（平台即服务）、SaaS（软件即服务）；在管理部分，主要以云管理为主，主要功能是确保有效的资源利用以及确保整个数据中心稳定、可靠、安全地运行。

云计算数据中心不仅仅包括服务器和与之配套的通信与存储系统，还包含冗余通信设备、监控设备以及各种安全设备。新一代的云计算数据中心针对成本、环保等问题进行改进，通过结合自动化、虚拟化以及新能源管理等技术打造与企业业务发展相适应的大规模低能耗绿色云计算数据中心。

传统数据中心主要是以面向设备为中心，而云计算数据中心的核心是面向服务，使用云计算技术将各种资源打包成服务，无论是设备、系统、应用等都是以服务的形态进行输出，并且可以随时随地交付给用户使用。

通常将云计算数据中心的参考架构分为以下五个部分：

基础设施：为了保障云计算数据中心的正常运行而建立的环境和服务设施，比如场地、网络、电力、制冷系统等设施。

资源池：是云计算数据中心的核心，是指各种硬件以及软件经过虚拟化技术进行整合的资源，一般分为网络资源、存储资源、计算资源、数据资源等。

安全管理：主要是针对云计算数据中心内部以及外部的各种安全威胁的防范措施，

保障云计算数据中心信息可以安全稳定地运行。

运维管理：主要是为了实现对云计算数据中心进行统一管理，包括运维管理机制、运维管理工具等，从而最大限度保障云计算数据中心各项功能正常。

能效管理：是指对云计算数据中心各种组件能源进行的检测和评估，包含能耗的计量、能耗评定以及管理机制。

8.2 云计算数据中心特点

云计算在某些方面具有云的特征，比如规模庞大边界模糊，可以动态伸缩，因此云计算数据中心的内容也是随时间的变化而动态变化的，会随着社会的不断发展进行优化与调整。但总体上云计算数据中心具有如下一些特点。

1. 高度虚拟化

云计算数据中心的首要特点就是高度虚拟化。要在有限的空间内部署更多的设备，如机架、刀片服务器和集装箱式数据中心等，而由于云计算数据中心内部服务器设备种类性能各不相同，只有将所有可以提供的物理资源和计算资源进行虚拟化整合才能够充分有效地加以利用。

常见的虚拟化技术包括服务器虚拟化、网络虚拟化、存储虚拟化以及应用等多种虚拟化技术，将一组服务器、网络以及存储通过虚拟化技术对外看做一台超级计算机、一个网络设备以及一台大型存储设备，通过提供简单的用户接口实现虚拟资源的快速创建和重新部署，达到对资源的统一和动态分配管理。

2. 高可用

为了保障对云计算用户提供安全、可靠的云数据服务，云计算数据中心的网络必须能够确保可靠的网络接入，任何一条链路失败都能由另外一条链路代替，实现永久在线的网络连接。并且数据中心中各个部分需要进行冗余设置，任何一台服务器出现故障，云计算管理平台都能够自动发现并且把出现故障的服务器从可用服务器列表中剔除，保证所有运算资源都能够建立在可用的服务器上，同时故障主机的计算资源能够自动迁移到其他可用服务器上，确保应用服务的不间断性。存储配置通过容灾系统确保数据的高可靠性。

3. 分布式存储

分布式存储系统是利用大量服务器资源来满足更多存储需求的有效方法，能够对存储资源进行统一管理，确保数据在整个读写过程中对安全、可靠等方面的要求。

最常见的是利用分布式文件系统实现的分布式存储技术，使数据可以分散存储在多台远程服务器上，在访问数据时像访问本地文件系统一样访问远程存储服务器上的数据。分布式文件系统都会设有冗余备份机制和容错机制来确保数据的安全性和读写正确性，云计算数据中心的存储服务基于分布式文件系统并根据云存储的特点做了相应改进，会对整个存储系统中的节点进行监控和检测，并有相应的故障恢复机制。

4. 自动化管理

由数据中心发展为云计算数据中心的另一大重要指标就是能够执行 7×24 小时无人值守的、完全的自动化运行，这里的自动化既包含对物理资源的自动化管理，也包括对虚拟服务器的自动化管理以及对相关业务的自动化管理。比如可以对服务系统漏洞进行统一的修复、快速部署各项服务、主动对系统性能进行分析管理等等。云计算数据中心对大部分常见故障进行监控，一般会在故障发生初期向管理人员发出报警，并且根据定义的机制进行自动故障处理，在一些特殊的情况发生初期，管理员也可以进行远程管理。

5. 绿色节能

据统计标明数据中心大部分能源消耗主要来源于电力系统，为了打造绿色节能的云计算数据中心,将大量使用节能设备，包括服务器、存储设备、供电系统、散热设备等，通过先进的供电和散热技术，再加上新能源技术的利用，以降低云计算数据中心耗能，最大程度提升能源利用率以及减少对环境造成的影响。

8.3 网络应用架构

云计算数据中心通过高速链路和交换机连接，包括从前端的客户端到服务器的网络以及后端的服务器端到存储端的网络，向用户提供高效的信息服务。在云计算模式下服务器和服务器之间交互数据的带宽需求量非常大，所以网络延迟问题成为一大瓶颈，使得更高的双向带宽需求更加迫切，以及尽量使用低延迟服务器也迫在眉睫。除了网络延时，还需要关注网络的收敛性，在二层使用 FibriePath 实现多路径协议，可以降低网络拓扑改变时的收敛时间，支持更大规模的无阻塞二层网络，减少大规模网络中生成树的影响，从而提高网络的可靠性，并且提升全网性能。

1. 拓扑设计

云计算数据中心网络关系到数据中心的整体架构。目前数据中心网络普遍采用树型结构的三层拓扑方案，最顶层是核心交换机，其次往下依次为汇聚层交换机和接入层交换机，但这种传统拓扑方案已经不能很好满足云计算数据中心的业务需求。为了解决传统三层拓扑方案中的不足，改进的新网络拓扑方案应运而生。

新网络拓扑方案分为以交换机为核心的拓扑方案和以服务器为核心的拓扑方案。

以交换机为核心的拓扑方案中，网络的连接以及路由功能是由交换机来完成的，常见方案包括 FatTree、VL2、OSA 等。

FatTree 网络拓扑结构仍然采用三层树型结构，与传统树型结构的三层拓扑方案不同，接入层交换机和汇聚层交换机被划分为不同的集群，在同一个集群中接入层交换机和汇聚层交换机都相连，而每个汇聚层交换机与某一部分核心交换机连接，如图 8.1

所示,这样的拓扑结构使得每个集群都能与任意一个核心交换机相连,有效解决了传统三层拓扑方案中存在的单点失效和带宽瓶颈问题。但 FatTree 网络结构的扩展会受限于核心层交换机。

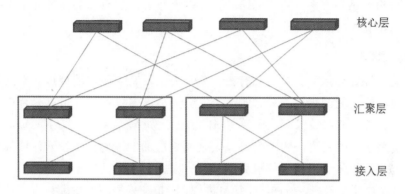

图 8.1　FatTree 网络结构

VL2 也是对传统三层拓扑结构进行改进,将多台服务器组成服务器组连接到一台接入交换机,每台接入交换机与两台汇聚交换机相连,每台汇聚交换机与所有的核心交换机连接,各级交换机之间都使用 10G 端口互连,如图 8.2 所示。是比较灵活的可扩展网络架构,但在服务器组和接入交换机间会面临单点失效的问题。

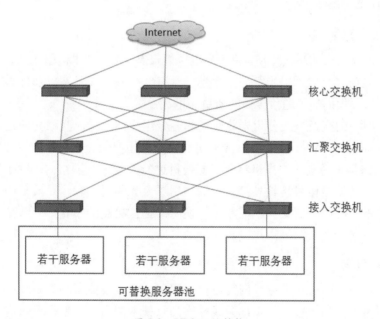

图 8.2　VL2 网络结构

以服务器为核心的拓扑方案中,网络的互连和路由功能主要由服务器来实现,交换机仅提供简单的交换功能,所以服务器往往有多个网络接口接入网络,常见方案包括 DCell、FiConn、BCube、MDCube 等。大部分方案具有层次性,比如第 0 层网络一般都是由一台交换机连接若干台服务器，高层网络是通过连接若干个低层网络构成。直接通过服务器端口进行连接的拓扑会受限于服务器端口的数量，所以交换机只出现在第 0 层，如方案 DCell 和 FiConn。由交换机进行相连的方案是 BCub 和 MDCube。采用这种结构的拓扑方案，对于网络的扩张性能较好，所有服务器都处于平行或并列的位置，比较节省网络架设成本。

2. 隔离技术

典型的云计算环境，无论是私有云、公有云都具有多租户的特点，不同用户的不同业务需要网络层面的隔离，只有实现了安全隔离，多种业务才可以整合在同一个物理网络上，传统的网络技术可以实现逻辑隔离，比如 VLAN、PVLAN、VSAN 等用来实现数据链路层面的隔离，VRF、VPN 实现网络层面的隔离。现在云计算数据中心的核心交换机上又提供了进一步的隔离方法——虚拟交换机技术，实现将一个交换机逻辑划分为多个虚拟交换机，与物理交换机类似，虚拟交换机主要由虚拟端口或端口组、上行链路和上行链路端口几部分组成，在进行数据传输时，虚拟交换机内部就是依靠这些端口或端口组进行的，与物理网络交换数据时才通过上行链路来进行。这些交换机之间是相互隔离的，都拥有独立的二层、三层协议，虚拟交换机之间是无法通过配置实现相互连通的，属于彻底的隔离，所以它们之间的环境是完全独立的，当某个虚拟交换机出现故障，也不会影响其他的虚拟交换机，可以实现网络资源的灵活部署与调度以及有效保障数据的安全。

3. 光交换网络

为了实现光信号的直接交换，免去光电转换限制，光子技术被引入到交换系统中。光交换网络的物理介质采用光纤，因为光纤具有低衰减、高带宽以及优秀的抗电磁干扰特性，从而大大提高了通信传输的质量和可靠性。

光交换网络的交换方式分为光路交换和光分组交换两种。光路交换是在主叫和被叫两个终端间建立一个光连接通道，光分组交换是采用分组交换技术的一种信息交换。

OSA 网络架构通过光交换机将所有交换机进行互连，在网络内部除了服务器与接入交换机之间使用电信号传输外，其余均采用了全光信号传输。是典型的应用在集装箱规模的数据中心的网络架构，依靠光交换网络提供更高的带宽，有效减少了传输过程中的信号衰减。

4. 无线网络

在云计算数据中心引入无线网络可以克服因大量突发流量和高负载服务器而降低数据中心网络性能的问题。特别是无线网络高频技术的引入，使得可以进行更高速度的数据传输，吞吐量可达到 4Gpbs。因此引入无线网络连接来分流有线链路的数据流，按需建立无线网络，使得数据中心网络可以动态调整拓扑结构，更适应当前网络环境，同时也降低了布线的复杂度，减少了数据中心网络建设成本，从而大幅度提升数据中心网络性能。

5. 软件定义网络

软件定义网络（Software Defined Network，SDN）是当下的新型网络架构，对网络进行抽象从而屏蔽复杂的底层网络环境，为用户提供一种简单的、可编程的软件模式来实现网络的自动化配置与管理。

SDN 网络架构分为基础设施层、控制层和应用层。基础设施层主要包含基础的交换机、路由器以及网络芯片等底层转发设备；控制层通过控制软件来集中管理控制不同的网络设备，包括数据流的接入、数据包的分析、路由机制、网络虚拟化等功能；应用层包含提供 SLA、QoE、安全与防火墙等网络服务。

SDN 与普通网络的最大区别在于改变了传统网络设备的转发与控制层的行为，将控制与转发有效分离，支持第三方控制设备通过 OpenFlow 等开放协议远程控制硬件的交换以及路由功能，还可以通过软件编程方式来满足客户定制化的需求，使网络的运行维护只需要通过软件进行即可。

OpenFlow 是 SDN 的核心技术，是 SDN 三层架构中的第一个标准通信接口，允许直接访问和控制网络设备的转发，从而实现了网络流量的灵活控制有效降低整体网络设计的复杂度以及网络构建成本。

8.4 能源利用

随着大规模云计算数据中心的广泛部署，高耗能、高污染等环境问题也随之而来，如何解决这些环境问题，实现节能减排，成为建立绿色云计算数据中心的研究重点。

绿色云计算数据中心最大的一个标志就是使用大量的虚拟化技术，充分调用所有现有硬件设备，高效利用虚拟化资源，因为可以对系统性能进行横向扩展，不必增加多余的服务器，从而减少硬件设备的投入，资源的能耗也相对减少，同样数据中心整体耗电量也会降低。

除此之外节能技术的投入，新能源的利用也是建立绿色云计算数据中心的关键。

1. 节能技术

云计算数据中心持续运转的重要环节是电力系统，而大部分的能源消耗在电力问题上，目前的电力能源主要还是各种化石燃料，所以节约电费成本、减少碳排放是首先需要解决的问题。

为了进一步改善云计算数据中心配电系统方案，降低电源系统建设维护成本，数据中心设备一般采用 240V 高压直流供电，利用机架分散供电能够将 240V 直流电直接转换为 12V 直流电，直接使用在多数标准交流设备上，可提供高达 92% 以上效率的供电。240V 高压直流系统内部的模块化设计，支持热插拔，维护方便，可以按照实际需求进行配置，随时增加或者更换掉出现故障的模块，那么就可以根据数据中心实际的负载情况设置自动开启所需模块，这对于数据中心有着明显的节能效果。

除了使用标准的 240V 高压直流供电之外，另一种各大互联网公司提出的高效节能技术是市电直供配电，市电直供配电保留了机架式电源支持热插拔的模块化设计，另外还最大化地减少了配电系统的转换环节，从市电到服务器只需要经过两级电路转换，将220V交流电提升至400V直流电，经过降压器转换为12V的直流电直接供给服务器使用。

目前大多的双电源服务器，正常工作时是两路服务器电源同时输出，自动均流各承担一半的负载，当其中一路电源出现故障，由另外一路承担全部负载从而保证设备的可靠供电。对于双电源服务器可以使用一路240V高压直流供电，另一路使用市电直供，服务器电源内部自动均流，两种供电方式均承担一半负载，这样可以使得综合供电效率提高。但如果考虑服务器电源容量的冗余，实际单个服务器电源的负载将会降低，电源转换过程会造成能源浪费。所以一般采用服务器两个电源模块互为备份的状态，主供电分路为市电直接供电，热备供电分路为240V高压直流供电，市电正常的情况下不使用热备供电方式，240V高压直流供电系统不承担负载，会一直处于热备状态，只有当市电出现故障，才会启用热备的240V高压直流为服务器进行供电，该方式对减少数据中心运行耗能有明显效果。

在数据中心，电力的中断意味着所有设备停止运行，所以在配电方案中都会标配UPS不间断电源，提供暂时的电力支持来保障网络设备的持久运行。相比传统的UPS解决方案，目前各大厂商推出了自己模块结构的高频UPS不间断电源，与传统UPS解决方案相比，高频UPS取消了谐波过滤器和输出变压器，本身功耗变小，单机UPS工作效率提升5%，还可实现在线热插拔和热更换，具有良好的扩展功能，UPS不间断电源根据自身设计，可实现断电后针对不同输入电源进行自动切换工作，同时对外干扰小，在节能减排、降低能耗方面具有优势。

2. 新能源应用

除了传统的供电方式，充分利用新能源供电也是重要的途径。新能源是指在传统能源（煤炭、石油、天然气）之外的各种可再生能源，又称非常规能源，包括太阳能、风能、生物质能等，一旦建设新能源电厂，就能源源不断地提供电能,管理费用相对较低。

在数据中心的整个电力系统的运行过程中，制冷设备能耗通常达到50%以上，制冷环节主要由空调负责，除此之外空调系统还为其他核心设备提供机房的加湿、除湿等功能。为提高空调系统运行效率，使用自然冷却系统成为关键。在一些海拔较高和高寒地区比较适合使用室外低温自然风或者低温冷水来为数据中心降温，比如Google在比利时的圣吉兰数据中心使用室外空气来进行降温，但是自然风会受到自然条件的影响，比如风速会影响发电量。完全的自然风冷却在我国应用不广泛，从而在有条件时采用自然风冷却，其他时候使用常规的空调制冷的混合模式应运而生。太阳能发电在数据中心的应用也比较广泛，但是太阳能会受到日照强度的影响。如果有丰富的自然低温水资源利用也是不错的选择，通过水泵驱动室外冷水循环达到冷却数据中心所产生的热量的效果，再由室内冷风冷却机房内的设备，带走设备运行时所产生的热量，从而降低数据中心电费。除了利用自然低温水资源，使用其他氟化液或者是特殊不导电的油来代替空气的液体冷却方法也得到应用,比如百度 M1 数据中心,对其数据中心进行水蓄冷改造以降低数据中心整体能耗。

目前国际上使用 PUE(Power Usage Effectiveness)即能源使用效率来衡量评价数据中心能源使用效率。PUE 的值等于数据中心消耗的所有能源与 IT 设备负载能耗之比(IT 设备负载能耗包括服务器设备、网络设备、存储设备以及用于管理、分析等外围设备的电量使用;数据中心消耗的所有能源除上述耗能外还包括主要的用电系统、备用电力系统、冷却系统、消防系统、监控系统、照明系统等基础设施的所有耗电量),比值基准为 2,越接近 1 表示数据中心的能源消耗小,主要任务是用来处理与分析数据,效能越高;接近 2 或高于 2 则表示在冷却系统等其他基础设施上有太多的能源消耗。在数据中心初期的规划中就需要把节能环保的问题考虑到位,并且在设计数据中心机房时把 PUE 值作为机房设计和规划的参考。

8.5 自动化管理

实现云计算数据中心重要的一步是自动化,云计算数据中心都是由成百上千甚至更多的服务器组成,那么要把分布在不同地理位置上的众多服务器资源集中调用进行协调工作,需要充分使用自动化的控制技术,达到对数据中心设备、网络、服务的统一管理,最大化地减少人工与设备的投入。

当数据中心资源实现虚拟化后,可以使用策略推送通过面向服务的方式实现数据中心的自动化部署和管理。对服务器的自动化管理,完成系统以及应用程序的批量部署,资源的自动部署和回收以及计算能力和存储能力的自动迁移,并对数据中心各种设备进行全面的性能监控,实时的数据采集,自动执行数据中心内的环境监测与修复,实现随时的故障设备自动迁移。对网络的自动化管理,可实现自动发现网络设备、拓扑变更,自动批量下发路由表和防火墙等安全策略,监控网络流量。对温度控制的自动化管理,通过安装部署热传感器,实时收集数据中心温度、湿度等环境数据,将收集到的数据传递给自动化温度控制平台,根据传递温度的设定值系统会自动调节冷风的速度和大小,从而也降低了能源消耗。除此之外,还可以自动分配和优化服务,比如对于 CPU、内存、磁盘、I/O 等资源设置系统阈值,这些资源一旦达到设定值,就会自动开启优化行为。

8.6 容灾系统

云计算数据中心的安全性不仅体现在防火墙、病毒防范、入侵检测系统等安全防范措施上,火灾、水灾以及地震等自然灾害也可能在任何时候对云计算数据中心造成破坏,在构建云计算数据中心初期就应该建立可靠的灾难恢复系统方案。

8.6.1 容灾概述

容灾系统是指在异地建立另一备份系统,相互之间可以进行健康状态监控和自动切换,利用地理上的特点降低自然灾害带来的损失。按照容灾系统对数据中心系统的保护程度,分为数据级容灾、应用级容灾和业务级容灾。数据级容灾是指异地的数据会被远程实时备份,在发生灾难时,只能确保关键应用数据不丢失,应用服务会在灾难发生时中断,数据级容灾系统构建相对简单,费用也是最低的;应用级容灾是指在异地建立的一套相同的应用备份系统,通过同步或者异步备份技术确保灾难发生时应用的快速切换,同时能够确保应用服务的不间断;业务级容灾是处理逻辑等非 IT 系统的冗余。

容灾系统的部署需要满足三个基本要素,也就是容灾的"3R"即 Redundance、Remote、Replication:首先是要确保数据中心中的部件设备、数据都具有冗余性,发生故障后能够确保数据的正常运行;然后是具有一定的距离,因为如果是自然灾害必定发生在一定范围的距离,只有确保足够的长的距离,才不会被同一灾害破坏;最后容灾系统需要进行全面的数据备份。除此,容灾系统的部署通常有两个衡量指标,RPO 和 RTO。RPO(Recovery Point Objective)即数据恢复点目标,表示业务数据所能容忍的数据丢失量,针对数据的丢失;RTO(Recovery Time Objective)即恢复时间目标,表示灾难发生后到业务系统恢复服务功能所需要花费的最短时间周期,针对的是服务。

针对不同的容灾需求,国际标准 SHARE78 将容灾系统划分为 7 个等级,如表 8-1 所示。

表 8-1 灾难恢复等级

等级	要求
0 级	无异地备份,未指定灾难恢复方案,不具备真正灾难恢复能力
1 级	异地备份关键数据(没有备份系统和网络)一般是将本地备份副本传输到异地保存
2 级	异地热备份(有备份系统无备份网络)
3 级	在线数据恢复,可通过网络将关键数据进行备份存放到异地
4 级	定时数据备份,备份数据采用自动化备份管理软件备份到异地
5 级	实时数据备份,两个站点之间互为镜像,实现双重在线存储,恢复数据可达秒级
6 级	零数据丢失,利用专用存储网络将关键数据同步镜像到备份中心,主备中心同时向外提供服务实现负载均衡

8.6.2 容灾技术

容灾备份技术作为保护云计算数据中心的最后一道屏障,是必不可少的环节。一般在建设云计算数据中心时都会建设两个或者多个数据中心,一个是主要云计算数据中心，主要用于承担用户的业务，另一个是备份云计算数据中心，主要用于数据、配置、业务等的备份。主备之间一般有热备、冷备、双活/多活三种备份方式。

热备和冷备都是其中一个数据中心处于运行中，另外一个数据中心处于不工作状态，只有当灾难发生导致数据中心业务瘫痪，灾备数据中心才启动运行。而双活/多活是两个数据中心均处于运行中，同时承担用户业务互为实时备份，也是云计算数据中心主要的备份方式。

以下是云计算数据中心采用双活进行备份使用的关键技术：

1. 镜像技术

在主备数据中心之间可以使用镜像技术实现容灾备份，按照镜像系统所处位置的不同，分为本地镜像和远程镜像，是容灾备份的核心技术。一般会在主数据中心建立的主镜像系统，备份数据中心建立的从镜像系统，即使灾难发生，分布在不同远程数据中心上维护的备份数据镜像不会受到影响，是保证数据中心之间进行同步和灾难恢复的基础。

2. 快照技术

快照技术是通过软件进行的备份，防止数据丢失的有效手段，并且能够进行在线的数据恢复。通常会把镜像技术与快照技术相结合来进行远程备份，一般先通过镜像技术把数据进行远程备份，然后使用快照技术把远程存储中的备份数据再备份到磁带库或光盘库中。

快照技术分为两类，指针型快照技术和空间型快照技术。指针型快照是把备份数据建立逻辑单元号 LUN 和快照 cache，将要修改的备份数据块复制到快照 cache 中，快照的逻辑单元号 LUN 是一组指针，指向快照 cache 和备份过程中没有改变的数据块，可以使用指针型快照对数据进行完整备份；空间型快照是在后台状态下为数据存储内部实时创建单独可寻址的多 COPY 卷，这些 COPY 卷存放当前生产环境的数据，备份系统可以使用 COPY 进行备份、开发测试等。

3. 远程复制技术

将数据进行远程复制需要满足数据的实时性、准确性、在线性几个方面。现在的存储设备尤其是中高端存储产品都是具有先进的数据管理功能，数据复制功能直接交给存储系统进行实现，利用 TCP/IP 网络将数据远程复制到其他的备份数据中心，典型的比如基于 IP 的 SAN 存储结构。

8.7 亚马逊 AWS

AWS 共提供 14 类 28 项服务，大致可分为计算、存储、应用架构、特定应用、管理这五大类，如图 8.3 所示。

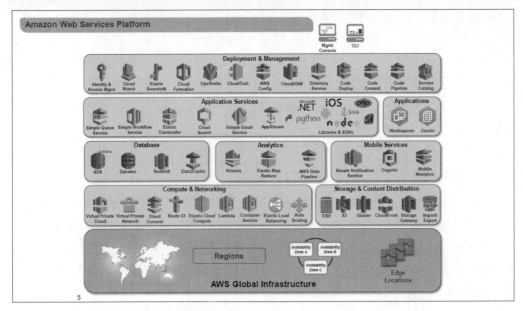

图 8.3　AWS 所有服务框架图

1. Amazon EC2 计算节点框架介绍

Amazon EC2 是一个 IaaS 云服务，主要提供弹性的计算资源。通俗的讲，就是提供多种类型的虚拟机。EC2 也是整个 AWS 最核心的组成部分，在 EC2 环境中，虚拟机被称为实例，实例的镜像被称为 AMI。

任何的企业和个人都可以选择不同类型和大小的实例，在很短的时间内创建、启动和运行，主要是根据类型和使用时间收费。

EC2 计算节点的主要功能如下。

- EBS（Elastic Block Storage）：为 EC2 实例提供永久性的存储。
- 弹性 IP 地址：用于动态云计算的静态 IP 地址。
- VPC（Virtual Private Cloud）：在 Amazon Web Services (AWS) 云中预配置出一个私有、隔离的部分，在自己定义的虚拟网络中启动 AWS 资源。
- Auto Scaling：可以根据定义的条件自动扩展 Amazon EC2 容量。

2. Amazon S3 服务

S3 全称叫做 Amazon Simple Storage Service，简单存储服务，是可扩展的云存储。

它是一个完全针对互联网的数据存储服务，应用程序可以通过一个简单的 Web 服务接口就可以通过互联网在任何时候访问 S3 上的数据。

S3 的数据存储结构简单，就是一个扁平化的两层结构：一层是存储桶（Bucket，又称存储段），另一层是存储对象（Object，又称数据元）。

更多关于 AWS 的介绍及基于 AWS 的大数据分析，请上课工场 APP 或官网 kgc.cn 观看视频。

本章总结

- 云计算数据中心在传统数据中心中引入云计算技术，实现了更大规模、更高集中程度与更加便捷的设备管理。
- 云计算数据中心的网络是关系到整个云计算数据中心的整体架构，常见网络拓扑方案有 FatTree、VL2、OSA、DCell、FiConn、BCube、MDCube 等。
- 云计算数据中心使用各种节能技术与新能源技术，旨在打造绿色节能减排的云计算数据中心。
- AWS 共提供 14 类 28 项服务，大致可分为计算、存储、应用架构、特定应用、管理这五大类。

随手笔记